Routledge Revivals

Metropolitanization and Public Services

Metropolitanization and Public Services is third in a series on the governance of metropolitan regions which aims to explore the welfare and development of Metropolitan America. Originally published in 1972, this study discusses who decides which essential public services need to be provided within a metropolitan area by looking at political reform as well as presenting ideas on functional efficiency, costs and benefits and the effectiveness of the political process. This title will be of interest to students of environmental studies.

Metropolitanization and Public Services

Charles M. Haar, John G. Wofford,
David L. Kirp, David K. Cohen,
Leonard J. Duhl, Edwin T. Haefele and
Allen V. Kneese

First published in 1972
by Resources for the Future, Inc.

This edition first published in 2016 by Routledge
2 Park Square, Milton Park, Abingdon, Oxon, OX14 4RN
and by Routledge
711 Third Avenue, New York, NY 10017

Routledge is an imprint of the Taylor & Francis Group, an informa business

© 1972, Resources for the Future, Inc.

All rights reserved. No part of this book may be reprinted or reproduced or utilised in any form or by any electronic, mechanical, or other means, now known or hereafter invented, including photocopying and recording, or in any information storage or retrieval system, without permission in writing from the publishers.

Publisher's Note
The publisher has gone to great lengths to ensure the quality of this reprint but points out that some imperfections in the original copies may be apparent.

Disclaimer
The publisher has made every effort to trace copyright holders and welcomes correspondence from those they have been unable to contact.

A Library of Congress record exists under LC control number: 74186472

ISBN 13: 978-1-138-12195-9 (hbk)
ISBN 13: 978-1-315-65065-4 (ebk)
ISBN 13: 978-1-138-12202-4 (pbk)

Metropolitanization and Public Services

NO. 3 IN A SERIES ON

The Governance of Metropolitan Regions

LOWDON WINGO, SERIES EDITOR

Distributed by
The Johns Hopkins University Press, Baltimore and London

Metropolitanization and Public Services

Introduction by

CHARLES M. HAAR

Papers by

JOHN G. WOFFORD

DAVID L. KIRP and DAVID K. COHEN

LEONARD J. DUHL

EDWIN T. HAEFELE and ALLEN V. KNEESE

Published by Resources for the Future, Inc.

RESOURCES FOR THE FUTURE, INC.
1755 Massachusetts Avenue, N.W., Washington, D.C. 20036

Board of Directors:
Erwin D. Canham, *Chairman*, Robert O. Anderson, Harrison Brown, Edward J. Cleary, Joseph L. Fisher, Luther H. Foster, F. Kenneth Hare, Charles J. Hitch, Charles F. Luce, Frank Pace, Jr., Emanuel R. Piore, Stanley H. Ruttenberg, Lauren K. Soth, Maurice F. Strong, Mrs. Robert J. Stuart, P. F. Watzek, Gilbert F. White.
Honorary Directors: Horace M. Albright, Reuben G. Gustavson, Hugh L. Keenleyside, Edward S. Mason, William S. Paley, Laurance S. Rockefeller, John W. Vanderwilt.

President: Joseph L. Fisher
Secretary-Treasurer: John E. Herbert

Resources for the Future is a nonprofit corporation for research and education in the development, conservation, and use of natural resources and the improvement of the quality of the environment. It was established in 1952 with the cooperation of the Ford Foundation. Part of the work of Resources for the Future is carried out by its resident staff; part is supported by grants to universities and other nonprofit organizations. Unless otherwise stated, interpretations and conclusions in RFF publications are those of the authors; the organization takes responsibility for the selection of significant subjects for study, the competence of the researchers, and their freedom of inquiry.

This study is the third in a series of papers resulting from an RFF-sponsored project conducted by an informal Commission on the Governance of Metropolitan Regions, chaired by Charles M. Haar of the Harvard Law School. Lowdon Wingo is director of RFF's program of regional and urban studies and a member of the commission. The papers were edited by Jane Lecht.

RFF editors: Henry Jarrett, Vera W. Dodds, Nora E. Roots, Tadd Fisher.

Copyright © 1972 by Resources for the Future, Inc., Washington, D.C.
All rights reserved
Manufactured in the United States of America
Library of Congress Catalog Card Number 74-186472

ISBN 0-8018-1392-1

Price $2.25

Contents

Foreword *vii*

Introduction *1*
CHARLES M. HAAR

1. Transportation and Metropolitan Governance   7
   JOHN G. WOFFORD

2. Education and Metropolitanism   *29*
   DAVID L. KIRP AND DAVID K. COHEN

3. A New Look at the Health Issue   *43*
   LEONARD J. DUHL

4. Residuals Management and Metropolitan Governance   *57*
   EDWIN T. HAEFELE AND ALLEN V. KNEESE

Commission on the Governance of Metropolitan Regions

Chairman
CHARLES M. HAAR
Professor of Law, Harvard Law School

GUTHRIE S. BIRKHEAD
Chairman, Department of Political Science, Maxwell School of Citizenship and Public Affairs, Syracuse University

ALAN K. CAMPBELL
Dean, Maxwell School of Citizenship and Public Affairs, Syracuse University

LISLE C. CARTER, JR.
Vice President for Social and Environmental Studies, Cornell University

WILLIAM G. COLMAN
Consultant, formerly Director of Advisory Commission on Intergovernmental Relations

FRANK FISHER
Consultant, Office of the Secretary, U.S. Department of Housing and Urban Development

JOSEPH L. FISHER
President, Resources for the Future

LYLE C. FITCH
President, Institute of Public Administration

BERNARD J. FRIEDEN
Director, Joint Center for Urban Studies of the Massachusetts Institute of Technology and Harvard University

EDWIN T. HAEFELE
Senior Research Associate, Resources for the Future

PHILIP HAMMER
President, Hammer, Greene, Siler Associates

MARK E. KEANE
Executive Director, International City Management Association

JULIUS MARGOLIS
Director, Fels Institute of State and Local Government, University of Pennsylvania

JOHN R. MEYER
President, National Bureau of Economic Research

ANNMARIE HAUCK WALSH
Institute of Public Administration

WILLIAM L. C. WHEATON
Dean, School of Environmental Design, University of California, Berkeley

LOWDON WINGO
Director, Regional and Urban Studies Program, Resources for the Future

Foreword

The papers in this volume and its companions are products of a long-standing interest of Resources for the Future in the welfare and development of metropolitan America. More particularly, they stem from an RFF-sponsored project that was launched in the spring of 1970 with the convening in Washington of an informal Commission on the Governance of Metropolitan Regions. Chaired by Charles M. Haar of the Harvard Law School, it is composed of scholars, practitioners, and experienced observers of the metropolitan scene.

"Governance" in the title—The Governance of Metropolitan Regions—is meant to imply something more than government. Webster defines it as "conduct, management, or behavior; manner of life" in addition to "method or system of government or regulation." In these papers, and those that will follow, the authors are concerned not only with the apparatus and process of government in the ordinary sense but also with the total interaction among people in their public capacities and interests, and between people and the public institutions. The dominating question is: How can the governance of metropolis be improved? And next: What must we learn to achieve this? Unless early progress is made in these directions the danger that hard-pressed American cities will crack under the multiple strains of old and new problems will be very real.

RFF did not embark on this effort expecting that metropolitan political reorganization would solve all metropolitan problems, but we are inclined to think that it will help, because we have seen so many obvious steps frustrated by the way in which history has organized our urban political life. Although the reform of the institutions of metropolitan governance is hardly a sufficient condition for the solution of major urban problems, our intuition is strong that it is a necessary one.

This, then, is the theme of the RFF project—the interrelationship of metropolitan problems and governmental structure. It has formed the basis of the deliberations of the Commission and has guided the preparation of the exploratory papers published in this series. The papers do not exhaust the issues of metropolitanism; their purpose is to add some dimensions to an already rich literature, some options for the policy makers. No blueprint for the future is presented, no definitive list of recommendations; the results hoped for are breadth of view, depth of perception at several critical places, and illumination of practical alternatives for action.

Many people contributed to this effort. Charles M. Haar, as chairman, negotiated the contribution of papers and materials. Lowdon Wingo first proposed the Commission as an effective exploratory device, administered the program for RFF, and oversaw the publication of these volumes. Daniel Wm. Fessler of the University of California Law School at Davis has been a valued advisor throughout. Professors Daniel M. Holland of M.I.T. and Karl Deutsch of Harvard made valuable contributions from their time and experience. Michael F. Brewer, while Vice President of RFF, was a faithful and effective participant in the enterprise from its inception. Add to these almost twenty authors and an equal number of Commission members and you have the elements for some new insights to the metropolitan problem.

Joseph L. Fisher
President, Resources for the Future, Inc.

Introduction: Metropolitanization and Public Services

Charles M. Haar*

Perhaps the most intractable cause of political strife and misgovernment is the division of the surface of the earth into political jurisdictions. While every new boundary drawn represents the solution to a political problem, once drawn it tends to become as fixed as the constellations in the night sky. The inertia of political boundaries is evident in the enormous energies necessary to alter them and the glacial pace of changes in the jurisdictions of settled political institutions. Social and economic conditions on the other hand can change very rapidly, so that there tends to be a cumulative deterioration in the "fit" of any set of governmental institutions with the needs and wants of the constituencies subject to their jurisdictions over time. In Africa national boundaries, which had their origins in the dim colonial past for reasons now virtually forgotten, do little more than exacerbate intertribal conflicts, as national governments pursue some workable political consensus in their diverse, fragmented national societies. Sudetenlands, Triestes, and Gibraltars have haunted Europe for centuries. The military pot has boiled for years in Asia and Latin America over boundary questions. In short, to draw an official line on a map is to define insiders and outsiders, and it is their interrelationship over that line which makes the problem. For a world of nation states, failure of governmental arrangements can produce widespread disaster.

But the world is also made up of lesser political entities each designed at some time in the past to resolve some social or economic problem. At this level the consequences of failure, taken case by case, are not quite so awesome. Misallocation of collective resources and the frustration of "grass

*Professor of Law, Harvard Law School.

roots" aspirations do not rend the fabric of society in quite the same way. But the costs of these cases can add up to periodic crises, the most recent of which have come to be known as the urban crisis in the United States.

Clearly this country suffers from such a misfit between actual conditions and political boundaries. Though this present inner tension results from evolution in economic, technological, and social developments rather than, say, the tribal loyalties of the third world, and though it has come more gently and less visibly over time, the cost in the mismanagement of resources and in the frustration of popular aspirations has been great. These papers on the governance of metropolitan regions constitute an attempt to redefine in a very pragmatic manner the essential issues posed by the lack of congruence between problem and political structure in the management of our great urban settlements.

Consider at the outset the great difficulties inherent in any attempt to alter established territorial allocations of political power. The enormous task of the American confederation and union was accomplished two hundred years ago only through the persistent struggle of a group of vigorous and relatively homogeneous men flush with newly won independence and united by a victorious struggle against a common enemy. The Constitution involved no redrawing of state or local lines: the primary questions before the Constitutional Convention had to do with the exercise of those powers and responsibilities that transcend local interests. The reform of metropolitan governance, however, addresses problems that tradition had considered to be of purely local concern—not questions of the proper size and training of the Army, or the intricacies of foreign relations, but those involving the neighborhood school, the family doctor, the local busline, the town dump, and segregation of the poor and black as it affects each man's home and neighbors, and each child's schoolmates. These are issues over which local disputes for control are often most bitter and intense, and on which many a liberal has discovered in himself a closet racist. They are issues that the reform of government in metropolitan regions must confront.

Because change in political structures is difficult, a strong case that it is necessary should be required of reformers. Such a pragmatic approach is a distinguishing attribute of the institutions and rules that guide collective actions, an attitude some impatient reformers criticize as mindless conservatism (though it does have an intrinsic tendency to err in that direction). After its hard-fought creation, we amend the Constitution not through majority vote but through a two-thirds vote of the Congress and the agreement of three-fourths of the state legislatures. A completed trial can be reopened by showing not that it might have come out differently but that it was marred by clear deprivation of rights or egregious errors in proceedings, or that new evidence has been found which is likely to have a material effect on the result. A rule of law is subject to challenge only when it can be demonstrated that the underlying rationale for it has ceased to exist. In the case for reform of the governance of metropolitan areas in our federal system, what is the new evidence? What are the changed conditions?

Cities are fundamentally machines to expedite interaction among men: to facilitate commerce, to nurture personal and communal growth, to satisfy instinctual drives, to advance art, and to make available the civilization that is a by-product of all of these. The needs persist, but the techniques through which they can be satisfied have changed rapidly in recent years. Our older great cities, which find themselves in increasing difficulty, are in the first instance the victims of progress, particularly in transportation and communication. New York, Baltimore, Philadelphia, Atlanta, Chicago, San Francisco, and Boston are all logically situated for a nation primarily dependent upon communications by rail and water. But no natural harbors are required by airplanes, and no permanent rails by trucks. The national economy is no longer physically bound to those booming central cities of the nineteenth century by rail, river, and canal. Yet these cities continue to dominate the physical development of the nation; major highway systems and airport locations established in the mid-twentieth century have been designed to serve the existing settlement pattern, not amend it. We now know that the creation of a large, functioning urban center where none existed before proceeds from a leap of imagination and an act of faith as much as from technological capability. New Town developers have amply documented this in recent years.

The apparent stability of the old settlement patterns, however, obscured for many years the fundamental changes that were occurring within and around those older cities. The most obvious impact of advances in communication and transportation technology was not in the rearrangement of the national settlement pattern but in the redistribution of economic and social activities and patterns within those areas of concentration. It is instructive to imagine a stroboscopic record of the growth of the major American cities over the last hundred years—a movie for example, made from a satellite camera hovering over each urban center taking a single photograph every day. Played back at 16 frames to the second, the city would grow before our eyes, each year condensed to 23 seconds. Streetcar lines would snake outward and subdivisions grow around them, then roads and highways and more houses; the growth would accelerate dramatically as the postwar years were reached, the explosive generative force of the central city would be made clear; and then, in the last few minutes, the decay of the center, as old and not-so-old buildings are abandoned and crumble into ruins without replacement, as the in-commutation patterns are replaced by out-commutation and suburban cross-commutation, as the racial composition of the central-city population shifts increasingly to black and brown. Occurring incrementally, these changes paced critical alterations in economic and social patterns only recently recognized. Now it has become infinitely more difficult to perceive the maladjustments as a whole, to identify and mobilize the individuals and communities detrimentally affected, and to develop and consider feasible responses.

The central cities, of course, have been the most evident casualties of suburbanization. As the demographic characteristics of urban areas continue to change, it is increasingly clear that the most affected include the elderly, the poor, the nonwhite, and the disadvantaged as well as all of those white middle-income families who prize city life but feel the pressure of the increasing tax burden, the diminishing level and quality of services, and the ominous possibility of violence. It is certainly difficult to believe that any "moral" solutions for this dilemma will emerge: the suburbs of each city are not likely to turn suddenly against the motive force that brought them into existence and throw open their subdivisions and their coffers to central city residents and mayors. Nor do the sovereign states appear interested in renouncing any more of their taxing or administrative powers to some new level of local government, a level that in many instances would cross state boundaries. On the contrary, one of the most important struggles in recent political history has been the fierce battle between the states and cities over control of the potentially vast flow of funds promised by federal revenue sharing. Indeed, the creation of metropolitan general governments like that of Miami-Dade County, appears to have relevance at best to isolated and special cases, and so is not likely to be the shape of political things to come. Are there paths yet unexplored?

A real hope for improvements in the governance of metropolitan areas may lie, paradoxically, in the possibility that the problems are greater than they have yet seemed—that they are so severe that unless they are dealt with on a metropolitan basis, and with the combined resources of the entire area coherently arrayed, the roof will fall in on everybody, urbanites and suburbanites alike. For the dominant underlying theme of the 1970s in America may well be the irresistible emergence of underlying interrelationships that have up to now been determinedly ignored. The famed skylines of many of our major cities are often as not lost in clouds of particulate matter and tenacious gases. With apocalyptic suddenness many of the most cherished features of American life, from the disposable soda bottle and massive automobile engines all the way to the flush toilet, are under indictment as contributors to a threat to the continued health of the biosphere upon which we all depend.

A great deal of unavoidable "new evidence" is becoming available in both physical and social terms, and in many different functional areas. It points directly to the need for a revamping of the distribution of responsibilities and powers within metropolitan areas to allow for unitary response at appropriate levels of scale to problems that are inherently indivisible—indivisible in the sense that they require just such a unified approach to make solutions even possible. It is at this functional level that the analysis of possible responses has been made in the essays that follow—not with any prior commitment to a specific governmental form, but rather with the pragmatic notion that form follows function and that the functions must therefore be reexamined in the light of the most current information on the nature of the problems with which they deal.

It is precisely on the level of such functional divisions that the greatest feasibility in the shifting of political boundaries has thus far been demonstrated, in the form of special districts, public authorities, and the like. But, in the forms tried thus far, the fragmentation of public power from general government to special units of government has presented problems of its own, which several of these essays address and which warrant political as well as professional analysis.

We are dealing here with four of the fundamental, nuts-and-bolts services that determine and define the nature and quality of our day-to-day life: education, transportation, health care, and residuals management. In the context of metropolitan areas faced with the need to deliver these services, the common questions are who should make the decisions, at what level, with what responsibilities, to which constituencies? But before considering the professional responses discussed in these articles it might be well to ask whether these are the right questions? Can we turn to the "procedural" problem of how and by whom the decisions will be made without first developing some consensus or professional assumptions about the substance? What do we want our children to learn? Where do we want to transport ourselves? How much health care, of what sort, do we require?

There is, of course, a chicken-and-egg aspect to this distinction, but it can help highlight a significant trend in the pendulum-like shifts of emphasis between recognition of the need for professional analysis and the need for political responsibility. For many years the balance has been on the side of technical skills: the use of neutral "experts," the emergence of the "technocracy" as a controlling force in both public and private sectors of the economy, the attempted exclusion of "political" considerations from local decision making. But all of these are coming under attack as implicit, but nonetheless distinct, values that have been identified in the purportedly neutral decisions on matters ranging from national transportation to local school administration. More and more we are aware that choosing the decision-making entity is frequently a de facto resolution of substantive questions, and at the very least results in a clear and significant emphasis in some definite direction—and that this very factor should therefore be considered in the first instance. One of the important services that the following essays perform is to help rend the veil of professed objectivity and expertise—allowing us to see how in each of the areas discussed the "who" and "what" of decision making are closely and often inextricably intertwined. In some areas, at least, it becomes clear that this is the central question that must now be faced.

These essays, then, are part of an overall effort to reopen the question of "who decides" on the essential services to be provided within metropolitan areas. The answer is subject to tests of functional efficiency, of the distribution of costs and benefits, and of the effectiveness of the political process. While some observers are less than sanguine on the benefits of restructuring decision making at a metropolitan level, others see much to be gained from innovations ranging from metropolitan-wide "vouchers" for services such as education to organizational "franchising" in public services. Jointly these essays provide a constructive contribution and impetus toward the rethinking of our political forms and concepts of the governance of metropolitan regions that is so critically needed at this time of swift and massive growth and change.

1 Transportation and Metropolitan Governance

JOHN G. WOFFORD*

It is not surprising that the recent report of the Committee for Economic Development (CED), *Reshaping Government in Metropolitan Areas* (1970), should have on its cover a photograph of a six-lane expressway, complete with cloverleaf interchange, passing through a thickly settled suburban residential area.

The photograph is significant in several respects. It conveys the importance of the private automobile in making possible the land-use pattern that dominates the metropolitan scene. It suggests the existence of major problems of governance connected with the planning, financing, and construction of major transportation facilities, including expressways. And most important, it suggests the critical importance of transportation facilities in determining the size and shape of our metropolitan areas.

To be sure, any one metropolitan problem is part of a "seamless web" of metropolitan problems; transportation is special, however, in that it literally *is* that web, or a major part of the web. Along with communications (telephone, radio, television) and delivery networks (gas lines, power lines, sewers, etc.), transportation is the major means of linking people or goods with other people or goods. As such, it is both a reflection of the kind and amount of activity at each end of a "trip," and a determinant of the location and level of that activity. Metropolitan transportation facilities are thus an essential physical structure of the metropolis itself, like the skeleton, or arteries, or veins—

---

*Director, Boston Transportation Planning Review, Commonwealth of Massachusetts. Formerly Director, Harvard Law School Urban Mass Transportation Study, and Executive Director, Governor Sargent's Task Force on Transportation.

7

the metaphor is best left imprecise. It is only natural that such an important part of a metropolitan area's physical structure should be intimately tied to issues of its governmental structure.

The first section of this paper will discuss the critical importance of transportation in defining what we call "metropolis." The second will describe the major metropolitan transportation problems, collected into categories that seem appropriate to the issues of governance. Next follows a discussion of ways *short* of changes in metropolitan governance that might help solve some of the major problems. And finally, the last section will suggest the role that changes in metropolitan governance could play in helping to solve transportation, and other, problems.

Transportation and the Boundaries of "Metropolis"

One can argue that transportation technology has been a major cause of "the metropolitan problem." How large, after all, is a "metropolitan area"? The Census gives us one kind of answer: a Standard Metropolitan Statistical Area (SMSA) is essentially a densely built-up core area (of at least 50,000 people), with contiguous areas of common interest. It is interesting to look at the way the Census defines the boundaries of those areas: "the entire population in and around a city, the activities of which form an integrated economic and social system...." The Census notes that the definition of an individual SMSA involves two considerations: first, the central city or cities; and second, "economic and social relationships with contiguous counties which are metropolitan in character...." The Census then defines the key word, "metropolitan": "The criteria of metropolitan character relate primarily to the attributes of the contiguous county as a place of work or as a home for a concentration of nonagricultural workers." And it then defines what is meant by "integration": "The criteria of integration relate primarily to the extent of economic and social communication between the outlying counties and central county." This is determined primarily by discovering whether a minimum percentage of people live and work in the same area.[1]

The layman's understanding of a metropolitan area is, I suggest, quite close to the Census definition, but perhaps more concrete. It focuses, I believe, on the transportation facilities of an urban area, and the way those facilities are used for commuting purposes. Such a layman's definition would sound something like this: *an area that includes at least one large central city and within which a substantial portion of people find it economical in terms of time and money both to live and to work.* If this is a valid definition, it helps to suggest the critical importance of transportation—particularly of commuting

[1] U.S. Bureau of the Census, *Census of the United States, 1960: Summary*, pp. xxvi, xxxi–xxxii.

patterns—in helping to define the subject under discussion: metropolitan governance.

The validity of this definition of a metropolitan area is reinforced by the notion that "urbanization" in this country is defined partly by the separation of one's place of residence from place of work. This was not always the case, and the change was significant. Such separation began early in the history of our cities, as Wade has said:

> Another and perhaps even more significant mark of increasing urbanization was the growing tendency of residents to work outside their homes. While settlements were still villages, most shopkeepers, tradesmen, and professional people lived and labored in the same place, using the front or first floor of the building for their calling and the rear or second floor for lodging. As towns grew, this practice broke down. Large numbers now sought employment in factories, mercantile firms, or construction projects. Rising rents forced many to move their families out of the shop into residential districts. Of all indices of urbanization probably none is more important than the separation of work from home.[2]

The layman's definition, reflecting this historical separation of work from home, also helps us to keep in mind the fact that the "metropolis" is a dynamic concept. As transportation technology makes it economically possible for people to choose to live (or work, or both) farther and farther away from the old core city, with no more than about 30 to 45 minutes from home to work, the boundaries of the metropolis are necessarily going to continue to expand, unless public policy intervenes to set limits on that freedom of choice.

Indeed, we have recently been reminded that transportation historically has been an important factor in determining both the physical and the governmental boundaries of the metropolis. Roger Starr has written of the critical role played by the Brooklyn Bridge in leading to the governmental expansion in 1898 of New York City from its smaller size (Manhattan and the Bronx) to an expanded city (including Brooklyn, Queens, and Staten Island). The bridge itself was completed in 1884, opening up large areas of Queens and Brooklyn to suburban development. Such development led to the need for more bridges. But with no federal or state aid for highways, and with the suburbs unable to afford the high costs of constructing spans across the difficult East River, New York City itself was the only political entity with the needed resources to pay for more bridges. And only by bringing the outlying areas into the city governmental structure could the city have an adequate return on its investment. Fourteen years after the opening of the

---

[2] Richard C. Wade, *The Urban Frontier: Pioneer Life in Early Pittsburgh, Lexington, Louisville, and St. Louis* (Chicago: University of Chicago Press, 1964), p. 308.

first bridge, consolidation thus seemed to be in the interest of both richer New York and the poorer suburbs. As Starr puts it, "without consolidation, no bridges; without bridges, no transportation to the city; without transportation, no land boom."[3]

There are obvious implications of continual physical growth (or "sprawl," depending upon your point of view) for issues of governance; suffice it to say here that it is transportation which has enabled the central city "to outgrow itself," and it is transportation which will do the same for the metropolis as we now know it.

Metropolitan Transportation Problems

It is easy enough to catalogue our metropolitan transportation problems. The CED report has done it recently and well.

> Most of the transportation systems of metropolitan areas were built decades ago and remain fixed in the same place. From the largest to the smallest metropolitan area, the grid pattern street network remains as jammed in the twentieth century as it was in the nineteenth. While improvements in traffic engineering techniques, such as one-way streets and reversible street lanes, have speeded vehicular traffic slightly, delays are still the rule rather than the exception.
>
> In the largest metropolitan areas especially, street traffic cannot be unknotted without large-scale construction of parking garages, basic changes in truck delivery schedules, and major investments in integrated systems of public transit. While the private automobile seems to be the cheapest and fastest means of personal transport in less populated sections of metropolitan areas, it is notoriously inefficient in central cities, particularly those with more than 750,000 in population. Yet it is unreasonable to expect travelers to use existing public transit facilities willingly in most central cities. Buses, which form the backbone of most public transit systems, are often old and uncomfortable, and are likely to follow inconvenient routes and schedules.
>
> As metropolitan areas like San Francisco and Washington are now realizing, improving public transit in the central cities will by itself do little to upgrade metropolitan transportation. More and more, both central cities and suburbs are accepting the fact that traffic delays and congestion can be avoided only by bringing together and meshing the plans and policies of state highway departments, airport and bridge authorities, rapid transit authorities, bus companies, and all local governments.[4]

---

[3] Roger Starr, "Power and Powerlessness in a Regional City," *The Public Interest*, no. 16 (Summer 1969), p. 6.

[4] *Reshaping Government in Metropolitan Areas* (New York: CED, February 1970), pp. 30–31.

There are a number of ways in which one can categorize the above set of symptoms:

(1) By mode of vehicle—cars, taxis, trucks, buses, minibuses, streetcars, subways, railroads, helicopters, jetplanes, STOL (short take-off and landing) planes, etc.

(2) By the kinds of facilities the vehicles use—expressways, arterial and surface streets, parking garages, steel rails, airports, etc.[5]

(3) By ownership of vehicles and facilities—public or private, individual or corporation, governmental agency or "independent authority," etc.

(4) By stages in transportation operation—planning, financing, constructing, operating, maintaining, etc.

(5) By the nature of the governmental units that have a role in various transportation functions—federal government agencies within the Department of Transportation (DOT), regulatory commissions, other cabinet departments, the Council on Environmental Quality, the Congress, Congressional committees, Office of Management and Budget, the Presidency; interstate bodies and compacts (river, port, and airport authorities); state government (highway departments, DOTs, independent authorities, state legislature, legislative committees, etc.); general metropolitan governmental bodies, as in Dade County, Toronto, Indianapolis, etc.; limited function agencies ("planning" agencies, transit authorities, etc.); municipal government (traffic department, public works department, urban renewal agency, hackney bureau, police department, budget office, mayor, city council, board of estimate, etc.).

(6) By an analysis of the informal, as distinct from the formal, "power structure"—highway lobbies; downtown business groups; suburban business groups; labor unions; neighborhood protest groups; "the transportation establishment" of bureaucrats, politicians, and private citizens; mayor's office; governor's office, etc.

None of the above approaches (and there are others) is particularly helpful to an analysis of the issues of governance. All either assume or ignore the very structural issues in metropolitan intergovernmental geography that must be faced directly as questions, not as givens.

The basic questions relate to power. What are its components? Who exercises what parts of it? With what connections to others and to the people? To what ends? With what degrees of effectiveness? In terms of transportation issues, the following questions and categories seem a useful way to approach these issues:

---

[5] It is worth noting, since it is often forgotten, that expressways can be used by buses as well as trucks and private cars—there is thus no necessary dichotomy between highways and "transit," where transit refers to all means of public transportation.

(1) In what ways do our transportation systems exhibit either *efficiency or inefficiency*?

(2) In what ways do our transportation systems exhibit either *equity or inequity*?

(3) In what ways do our transportation systems exhibit either *legitimacy or illegitimacy*?

"Efficiency" refers to issues of speed, comfort, convenience, safety, etc., chiefly from the point of view of the user. Does the system, judged by transportation criteria only, work reasonably well?

"Equity" means issues of the impact of transportation upon different individuals and groups, whether they use the system or not. Who benefits? Who bears the costs? Is there a fair balance?

"Legitimacy" encompasses issues of accountability, concern for the public weal by procedures that produce both public acceptability and ultimate public responsibility. Who makes the decisions? By what standards and what procedures are those decisions judged and constrained? What are the limitations of power?

Issues of Efficiency and Inefficiency

In a number of ways our metropolitan transportation systems show a good deal of efficiency. Free-flowing expressways, for example, are safer, speedier, and generally less polluting means of moving large numbers of people than were most existing arterial street networks. Similarly, the private automobile is generally an efficient means of moving people with extremely diverse points of origin and destination, particularly in small and medium-sized metropolitan areas. Buses, the prevailing means of public transportation across the country, are generally more efficient than streetcars and commuter railways for people with diverse origins and destinations. Furthermore, existing technology can make such existing facilities even more efficient. Reserved lanes for buses, metering devices to maintain "free flow" on expressways, TV monitoring of particular bottlenecks—all suggest that increased efficiency is both possible and probable.

These efficiencies relate only to parts of a system, however. The issue is whether there is systemwide efficiency. The clear answer is that there is not.

For example, even within the private automobile mode, there is enormous inefficiency: in most highway planning, there is virtually no consideration given to the critical relationship between expressways, interchanges, local street capacity, and parking policy. Decisions are made to extend expressways into metropolitan areas without any assurance that the downtown distribution and parking problems will be met, thereby accentuating systemwide discontinuities by moving bottlenecks from one place to another and even worsening congestion by attracting more cars to the highways. Furthermore, inadequate attention has been paid to issues of scale—e.g., how many lanes should expressways in metropolitan areas have? The answer from most state highway departments is to build them to the "maximum" width to handle as much of the peak commuter traffic as possible, which generally means four lanes in each direction, since five or more are considered significantly less safe.

If there are inefficiencies in the private auto mode, there obviously are inefficiencies in bus and subway modes—unreliable schedules, danger of crime, dirty and dilapidated cars, increasing costs, declining ridership.

The wasted time spent getting to and from our metropolitan airports, as well as the time waiting to take off, land, get to a gate, and wait for luggage suggest drastic inefficiencies in the air traffic mode.

Of greater significance for purposes of this discussion than inefficiencies within modes are the inefficiencies between modes. Critical "mode-mixing" problems are virtually ignored—park and ride facilities are generally designed for a difficult rather than an easy change from one mode to the other; downtown-to-airport facilities are notoriously time-consuming; downtown distribution systems of traditional public transit are frequently little more than a bus every 15 or 30 minutes up and down Main Street and several key arteries, usually as the inner part of a line-haul operation rather than as a specially planned distribution system.

Equally important in terms of the absence of system integration is the inattention in public policy terms to issues of staging—whether to build one kind of transportation facility before or after others (e.g., highways before rail transit extensions along given corridors, radials before belt highways, expressways before local street and parking improvements). Such staging decisions often have vital effects on long-term travel and land-use patterns; and they can have serious intermediate consequences in terms of congestion and harm to the surrounding environment. Yet such staging decisions are generally made only indirectly; the agencies controlling different modes or different aspects of the same mode race to be first, and the one with the most available funds generally wins.

There is not, in short, what one could reasonably call a metropolitan transportation system. There are a series of services and facilities, mainly disconnected, but an integrated system there is not.[6]

Issues of Equity and Inequity

Is there a reasonably fair balance between the benefits and burdens of metropolitan transportation facilities, services, and policies? The answer, I

---

[6]The above analysis has drawn heavily on Parts I and II of the report of the Massachusetts Governor's Task Force on Transportation (Boston, January and June 1970), of which the author was executive director.

submit, is no. Inequities abound. Some people receive vastly more benefits; others are dangerously and intolerably burdened. This is true both of direct costs and benefits, and of indirect costs and benefits.

In terms of direct costs and benefits, the peak hour suburban commuter tends to receive substantially more than he pays whether he is traveling on an interstate expressway, whose inner-city width was primarily determined by the level of peak-hour demand, or on a commuter rail or transit extension, the outer reaches of which are generally subsidized by the more densely patronized inner portions. The people who live in these inner portions are well known to be less affluent, less able to bear the cost. The subsidies that exist, in other words, are regressive in welfare terms.[7]

In terms of indirect costs and benefits, there are substantial unfair burdens placed upon three groups: those who do not have access to an automobile; those in the path of or adjacent to major transportation facilities; and those yet unborn.

There are a number of groups in metropolitan populations who either cannot or should not drive—the young, the old, the handicapped, the very poor—yet they suffer acutely from the lack of an automobile. This is so largely because our automobile-dominated society has substantially curtailed public transit service as more people have come to own and use cars, and because the land-use patterns the automobile has made possible require (in all but the dense inner-core area) transportation by means of a relatively small, rubber-wheeled vehicle which can travel both on neighborhood streets and expressways. It is only fair to ask the automobile-owning majority to pay for a substantial part of this indirect cost it has created. At present, such cost is borne by the immobile and by the general taxpayer.[8]

Those in the path of or adjacent to major transportation facilities also bear an undue share of the cost of enabling the users of those facilities to enjoy them. Relocation assistance and "just compensation" are inadequate to compensate even the displaced family; much less do they account for numerous other deleterious side effects of transportation facilities: destruction of part of the low-income housing supply; disruption of neighborhoods; congestion on surface streets around major expressway interchanges; jet noise interfering with sleep, education, and social discourse; noise and air pollution from autos, trucks, buses, subways, with impacts on life and health that are yet undetermined; destruction of parks and other open space; changes in neighborhood land-use patterns; impact on local tax base and employment. The

---

[7]See Martin Wohl, "The Urban Transportation Problem: A Brief Analysis of Our Objectives and the Prospects for Current Proposals" (The Urban Institute, Washington, D.C., Working Paper 450-6, March 2, 1970).

[8]See Alan Altshuler, "Transit Subsidies: By Whom, For Whom?" *Journal of the American Institute of Planners*, March 1969.

problem is not simply that such by-products of transportation investment are not compensated, which is bad enough; even more significant in trying to lessen such deleterious impacts is the fact that the range of nontransportation values implied in the above list is simply not taken into account in making major transportation decisions in the first place. Occasionally, "design" revisions attempt to ameliorate some of the problems, but such ameliorations come too late in the decision-making process to be very effective. Afterthoughts are not a satisfactory way to deal with such critical problems.

Finally, it was once assumed that such short-run deleterious consequences would be offset in the long run by beneficial possibilities for development and positive changes in land use, but this can no longer be accepted without question. Expressways that were originally justified as essential to the health of the downtown have been shown to be a major force pulling jobs and customers *away* from downtown. I for one do not particularly mourn the passing of the "central city," as it once was known.[9] I do, however, believe the varieties of life style that a metropolitan area possesses should be preserved so that people have the option of choosing a dwelling other than a one-family house on a suburban plot. Many transportation facilities now are seen not to have widened those options but to have narrowed them.

This is particularly true in regard to options open to future generations. Our transportation agencies have built and built as if resources of money and space were unlimited, much in the manner of transcontinental railroads crossing the plains. We are learning that with a growing population, our resources (clean air, open space, water, etc.), are far from unlimited, especially if we look 20 or 30 years ahead to the turn of the century. In terms of "equity," we are discovering that we have been borrowing from the future things that we cannot pay back. The destruction of irreplaceable commodities must be viewed with alarm.

Metropolitan transportation thus tends to be inequitable—to the immobile, to the poor, to those who live near major facilities, and to the future.

Issues of Legitimacy and Illegitimacy

Most major metropolitan transportation decisions are made in an illegitimate manner. The agencies making the decisions usually have a statutory mandate to do what they do; however, one could make a strong case that many of them have followed investment opportunities in fields only remotely connected to the public objectives that originally called for their creation. The point is that most major agency actions, however "legal" in a technical sense, do not accord with our notions of a government ultimately account-

---

[9] See David L. Birch, *The Economic Future of City and Suburb* (New York: Committee for Economic Development, 1970), Supplementary Paper No. 30.

able to the people. Legitimacy in democratic theory means accountability, and accountability means politics.

"Independent" authorities, of course, were established largely to be free from political domination and from public budgetary allocation processes. No longer would bridge, tunnel, port, and turnpike authorities have to vie with other claims on the public purse; they could go directly to bondholders, in whose interests they were supposed to act. In theory, the authorities are ultimately accountable to the people by the fact that their boards of directors are appointed by mayors, or governors, or other elected officials; but the terms tend to be staggered, and not of the same duration as those of the official appointing them. Effective control by elected political officials, therefore, depends solely upon informal ties between the boards and the officials. And aside from control, formal accountability—and accountability is partly a formal matter—is extremely limited. Annual reports, occasional veto provisions, and the need for some authorities to seek legislative approval for major new projects are about the extent of the techniques for accountability.

Even line agencies concerned with transportation are generally subject to only limited control from elected officials. These agencies, like the independent authorities, are frequently administered by boards of commissioners, with staggered terms that tend to be of different durations from the officials'. Mainly, of course, we are concerned here with state agencies, especially state highway agencies. Most state governors do not have the staff to provide control of these limited-purpose agencies, if indeed they have the legal authority to do so. Frequently, the agency head was not even appointed by the incumbent governor. To be sure, the agency budget is approved by the state legislature and the governor, but even here there is limited accountability; federal aid policies, which channel substantial federal funds into state transportation agencies, make the state-dollars involved an inadequate reflection of the full impact of the total transportation investment, counting both state and federal funds. And legislatures generally are not asked for project-by-project approval. It is not surprising that state legislatures and governors in the past have given less than rigorous attention to the activities of their transportation-building agencies.

Aside from limited accountability to the top elected official of the state, there is even less accountability to the people. Certain formal procedural requirements, such as hearings, are designed to make sure that people affected by transportation facilities have a right to be heard before a decision is made to build. Today, such hearings tend to have little impact: they are conducted by the agency, which has already made up its mind and which has the authority, subject to federal approval, to do as it wishes; citizens' arguments tend to be ignored. The hearings, conceived as a means of obtaining citizen input into transportation decisions and of providing an aura of ultimate acceptability even among groups opposed to the construction of a particular facility, tend in fact to exacerbate the situation and to expose to the public the extremely limited control it really has over such decisions.

Nor is limited direct input by citizens balanced by effective input or control by their local governments. Local governments have even less say over the independent authorities than do governors. And state highway agencies in most states have close to unlimited formal power to force expressways through unwilling municipalities. Local governments do tend to have effective control over such matters as arterial streets, parking policy, the regulation of taxis, the control of bus routes. In a few places, they have, in addition, veto power over the location of state-financed roads. But by and large, their role in major transportation construction projects is limited to that mandated under federal law: the federal Secretary of Transportation shall not approve any projects in an urban area of more than 50,000 people "unless he finds that such projects are based on a *continuing comprehensive* transportation planning process carried on *cooperatively* by States and local communities...."[10] This is not the place for a review of the results of this statutory mandate. Suffice it to say that some states have turned to regional and metropolitan planning agencies to fulfill the mandate, and others have established special state bureaus that have put more emphasis on the words "continuing" and "comprehensive" than they have on the word "cooperatively." To put the matter another way: in many places, the mandate has produced more emphasis on long-range master planning, and virtually none on creating mechanisms for truly cooperative planning that might make the plans themselves legitimate expressions of an ultimately democratic process.

These then are the major problems of metropolitan transportation. The system is inefficient. It is inequitable. And it is illegitimate. The next section will consider what role changes in policy and structure—but not in metropolitan governance—might play in helping to solve these problems.

Problems, Causes, and Non-Metro Solutions

For analytical purposes, this section will discuss ways to solve transportation problems without attempting any major governmental revision of metropolitan areas and without attempting any major reallocation of functions among existing levels of government.

First, though, let us note the variety of diagnoses one can make about the "causes" of inefficiency, inequity, and illegitimacy:

(1) Federal aid policy, which biases local decision making in two ways—

---

[10]Federal-Aid Highway Act of 1962, Public Law 87-866, 87th Cong., 2 sess. (1962), 23 U.S.C. 134.

first, toward highway investments, by making aid available in large quantities on a 90-10 basis; and second, toward investment in new capital facilities instead of in maintenance of existing facilities.

(2) State constitutional restrictions on the use of funds raised from gas taxes, bridge tolls, etc., thereby preventing effective intermodal financing.

(3) Lack of effective means at either state or metropolitan levels of coordinating the activities of various single-purpose transportation agencies.

(4) The habits of the transportation technicians to be jealous of their power, resistant to political control, reliant on long-range master planning without much attention to intermediate problems or to future implications of present decisions, suspicious of citizen involvement, willing to achieve "interagency cooperation" by a process of back scratching and horse trading among independent fiefdoms.

(5) Ineffective administration at state and metropolitan levels in enforcing existing federal and state procedural requirements to achieve maximum possible efficiency, equity, and legitimacy—instead of the minimum that the technicians prefer.

(6) State laws that limit the control of elected officials over key transportation agencies.

(7) Inadequate transit management reinforced by anachronistic work rules and labor agreements, which inhibit the introduction of technological innovations.

(8) Consumer preference for single-vehicle, door-to-door transportation—i.e., the automobile—coupled with the myth that such generalized preference automatically determines the location, scale, and scope of urban expressways.

There are obviously a number of ways to deal with these problems—without assuming either that changes in metropolitan governance are required or that such changes, even if introduced, would solve all aspects of the problem:

(1) A change in federal aid policy so that more funds are available for transit than at present and so that funding formulas for highways and transit are identical. Local, metropolitan, or state priorities could then be determined without significant influence from federal dollars.

(2) A change in the requirements of federal highway policy so that more federal aid could be made available for urban arterial street improvements and "mini-expressways."

(3) Creation of a federal Transportation Trust Fund, in place of the present Highway Trust Fund, to collect various federal user taxes and disburse them to states and metropolitan areas for intermodal use without regard to the modal source.

(4) A similar Transportation Trust Fund at the state level. Maryland has recently enacted a twofold trust fund, one financed from unrestricted

sources, the other from bonds held by "independent" authorities with a kind of "authority trust fund" disbursing the revenues.

(5) Vigorous state administration, generally by the governor, to set a policy and, it is hoped, to force compliance by the technicians at the state level.

(6) The same done at the municipal level by the mayor.

(7) Transit operations improved by getting labor and management jointly to establish an atmosphere of working together in the public interest.

(8) Enactment of federal and state provisions requiring that replacement housing be built before the construction of transportation facilities can demolish existing housing.

(9) Vigorous enforcement of section 4 (f) of the Highway Act of 1968, which tightened up on the findings required before parklands can be taken—namely, that "(1) there is no feasible and prudent alternative to the use of such land, and (2) such program includes all possible planning to minimize harm to such park, recreational area, wildlife and waterfowl refuge, or historic site resulting from such use."[11]

(10) Enactment of state laws stipulating that the heads of state transportation agencies be appointed by the governor, and serve at his pleasure (or at least for terms coterminous with the governor's).

(11) Similar municipal ordinances in regard to the mayor's power to appoint municipal transportation officials.

The list could continue. It only suggests the variety of approaches that are needed to begin to close in on the problems of "governance" that remain. For some do remain.

The changes in "governance" now to be described still do not amount to changes in *metropolitan* governance. Indeed, they explicitly reject the notion of any new metropolitan entity, or of any reallocation of functions among existing levels of government.

The kingpin of most current transportation proposals is the establishment of a strong state department of transportation. This is designed to achieve *efficiency* by providing interagency and intermodal coordination; *equity* by ensuring that major transportation policies go through a kind of "clearing house" review by other state departments with nontransportation values; and *legitimacy* by, essentially, considering the governor as the only elected official responsive to a metropolitan-wide constituency. As Doig has pointed out:

> The limitations of local governments and independent authorities often give state institutions significant responsibilities in coping with major re-

---

[11] The recent decision of the U.S. Supreme Court in *Citizens to Preserve Overton Parky. v. Volpe* (401 U.S. 402 [1971]) is an important step in mandating such vigorous enforcement.

gional problems. The governor's role is especially important, at least potentially. With his constituency increasingly metropolitan, the governor is likely to be responsive to metropolitan needs.[12]

Doig goes on to quote from Robert Wood's unpublished Ph.D. dissertation, which noted: "The Governor knows that the crucial vote lies within the city and its suburbs."[13] Wood's thesis, written at Harvard University in 1949, was called "The Metropolitan Governor."

The metropolitan governor can be seen as, in effect, the mayor of metropolis. Letting the governor function in this capacity may work fairly well where the state capital is also the central city of the state's largest metropolitan area, so that the people of the metropolis feel that they have easy access to the State House. This situation obtains, however, only in the following eighteen states: Arizona, Arkansas, Colorado, Connecticut, Georgia, Hawaii, Idaho, Indiana, Iowa, Massachusetts, Minnesota, Mississippi, Oklahoma, Rhode Island, South Carolina, Utah, West Virginia, Wyoming. Not on the list are most of the large industrial states, and most major metropolitan areas. My surmise is that the governors of such states as Pennsylvania, Florida, California, New York—each of which has a state department of transportation—do not, and would not, attempt to serve as the mayor of their major metropolitan areas.

In Massachusetts, where the governor does act as a kind of mayor of Greater Boston, the Governor's Transportation Task Force recently recommended not only that there be a strong Secretary of Transportation, with the ultimate power to make and carry out decisions, but also that additional steps be taken to deal more completely with what in this paper are called the problems of equity and legitimacy.

To achieve multivalue planning and citizen input, the Task Force recommended that a strong Secretary of Transportation be balanced by "countervailing planning capabilities spread throughout the interested parts of the political system—including the modal transportation agencies, the appropriate regional planning agency, local planning departments, and nongovernment groups—to permit widespread informed participation" in the making of major transportation decisions.[14] To implement this recommendation, the Task Force proposed improving local planning capabilities by providing technical assistance in various ways; providing easy and timely access to official data through a state "Freedom of Information Act"; and creating widely varied forums for constructive participation in the planning process, including a

---

[12] Jameson W. Doig, *Metropolitan Transportation Politics and the New York Region* (New York: Columbia University Press, 1966), p. 8.

[13] *Ibid.*, p. 253.

[14] Governor's Task Force on Transportation, "Report to Governor Sargent, Part II" (Boston, June 1970), p. 6.

variety of working committees—composed of state agencies, local governments, and private groups—to study controversial plans and corridors.

Also proposed by the Task Force as a balance to a strong Secretary was the creation of "fair and formal procedures . . . to guarantee that all significant viewpoints receive a serious hearing" (p. 6). To implement this recommendation, the Task Force proposed that the state Department of Transportation support each proposal for a new transportation facility with a rigorous set of findings, including an analysis of alternatives, and the costs and consequences of each; and that these findings be subjected to an open public hearing, conducted under the auspices of a quasi-independent review board, where an adversary proceeding could be used to evaluate the findings. These recommendations the Task Force called "procedural due process" (p. 15).

In discussing the relationship among a strong Secretary, countervailing planning capabilities, and procedural due process, the Task Force declared:

> We view these components of the decision process as complementary. No one can be omitted without seriously impairing the entire process. In the absence of countervailing planning capabilities, the availability of information would have little significance. In the absence of open procedures, administered with the aim of producing meaningful dialogue in an atmosphere of trust, countervailing capabilities might simply produce an intensification of conflict. In the absence of a strong Secretary, the process would be highly susceptible to paralysis (p. 6).

This threefold set of recommendations is an attempt to establish efficiency (the strong Secretary); equity (participatory planning and multivalue findings, and ultimate decision by the best balancer of metropolitan interests, the Governor through his Secretary); and legitimacy (participation by local governments in the planning process; reasoned discourse on the findings; the chance to be heard in the independent hearings; and accountability through the Governor). The hope is that this process will create a good deal of consensus, and when consensus is impossible, at least the differences will be aired.

We are now attempting to implement these recommendations with respect to certain major transportation controversies. However, some of the following problems with the model can be expected:

(1) Who represents each municipality in the participatory planning process? The mayor? The city council? The mayor and council? The planning department? The state representative(s) and senator(s)? The U.S. Congressman? Some of these may well have significantly different views, which in turn represent differences within the municipality. Some minority viewpoints may not be represented at official levels at all.

(2) What does one do about changes in local views over time? For example, the replacement of one mayor (prohighway) with another mayor (antihighway). One certainly wants the process to be flexible enough to absorb

such changes in view, especially if the highway was a major issue in the mayoralty election, but this does create problems in consensus building. This problem is especially relevant in view of the many years between first plans and final completion of new facilities.

(3) Can technical assistance really be provided by the state so that local groups and communities can develop alternative plans with which to evaluate those put forth by the state agency? Technical assistance can be a tool of cooptation. It can also be a cloak for purely destructive community organization. How to keep it vigorous yet constructive is the challenge.

(4) Whom do the private groups really represent? How can we have any confidence that we have tapped the major differences in views, values, priorities, preferences? Many private groups will not show any interest until they hear that a particular plan might affect their neighborhood. The process must be open to such late entries.

In addition to these difficult procedural questions, we must also face the fact that there are deficiencies in the model in terms of reaching our objectives of efficiency, equity, and legitimacy.

In terms of efficiency, there is still limited means to coordinate expressway policy with those other key elements of the highway system—interchanges, arterial street improvements, parking facilities, and prices. All of these except interchanges are substantially controlled by municipalities, with little need to coordinate with state plans. Effective coordination is possible, but it is not automatic. Even more important, so long as each municipality needs to develop its own tax base, each will feel the need for some kind of expressway or transit access to an industrial park—even if more rational metropolitanwide policy would determine the location of such employment centers without considering the issue of local tax base as such an important constraint. Finally, still by way of illustration, the licensing and regulation of taxis—which can be a key component of transportation policy—is still almost totally within the control of each separate municipality.

In terms of equity, it is not at all clear that the combination of a strong Secretary, participatory planning, and fair and formal procedures really provides the kind of forum needed for a fully effective balancing of interests. In the model suggested, those interests get balanced that are alert enough to be interested. This may, of course, always be the case in politics, but one likes to think that the availability of a forum for the recognition and reconciling of interests will itself create an interest in being recognized and reconciled. In other words, a participatory *transportation* process may be too narrowly focused to get the input of citizens who are not primarily concerned with transportation but would be concerned about some of the impacts of transportation on other interests and priorities if they had a forum with a multi-value focus.

And in terms of legitimacy, there is a real sense of fuzziness about which private groups and which municipalities one should seek out in order to obtain a fairly accurate cross-section of views. Whom do the private groups—the neighborhood protest groups, the chambers of commerce, and others—really represent? Generally, they represent the activists in their midst, either at the staff or executive-committee level. This can be an insufficient mechanism for assuring legitimacy. Similarly, in terms of municipalities, should one give a greater role in the planning process to those municipalities through which proposed transportation facilities would pass, as opposed to those municipalities containing inhabitants who would *use* the facilities? Clearly, both should be adequately represented, but those feeling a direct impact are more likely to play an active role in the planning, if they are permitted to, than those who would like to use the facility passing through some other locality. And in either event, if a decision is ultimately against the interest of the particular municipality, it is not clear that the relative informality of the participatory planning process will provide the kind of ultimate aura of acceptability that we seek in trying to make our governments accountable to the people. Ultimate accountability through the governor to the entire state electorate may or may not satisfy the metropolitan constituency.

This raises, of course, the overriding uncertainty of politics. In relying upon the governor as mayor of metropolis, what happens if an antimetropolitan governor is elected? What happens if there is a major fight between the central city mayor and the governor? What if a particular governor is strongly pro-transit and anti-highway, or pro-bus and anti-rail, or vice versa? Enormous power has been concentrated at the top. Will the countervailing planning capabilities and the provisions for procedural due process act as an effective balance without leading to paralysis? It is too early to say.

This section has noted that there are many governmental solutions to metropolitan transportation problems that are not solutions of *governance*, and that some of these changes in law and policy are going to be required even if major changes in governance occur. This section has also tried to establish that one can make major changes in governance—the way institutions and interests relate to each other in a political decision-making process—without either creating a new metropolitan entity or reallocating functions among different levels of government. The model proposed above is not entirely satisfactory, but at least it is a step in the right direction, assuming that an incremental approach is the most likely to be implemented.

An Attempt at Metropolitan Constitution Making

The model of a metropolitan transportation decision-making process just discussed is an attempt to go beyond the current trend toward reliance on a strong state Department of Transportation, and—with participatory planning and fair and formal procedures—to make the entire process more efficient, more equitable, and more legitimate. It is worth noting, however, that the Massachusetts Task Force, which recommended trying out the above model, declared:

> We note explicitly, and with regret, that our recommendations assume the continued absence of strong metropolitan government. If such a government existed—with a strong executive and a representative, coherently organized, legislature—we might have urged that it be granted a veto over transportation projects in the area of its jurisdiction. We have judged that according such a veto to any of the existing regional boards or councils, however, would result in stalemate and stagnation.[15]

This section will discuss the possibilities of creating a new metropolitan governmental entity, and the role that such an entity would play in solving major transportation problems facing our metropolitan areas.

It is clear that one role for a metropolitan governmental entity could be simply to take over from the state on the one hand, and from the municipalities on the other, various transportation functions. Such a functional reallocation to a metropolitan entity has been done in both Dade County and Toronto. In Dade County, for example, the first three enumerated powers of the new Board of County Commissioners relate to transportation: power to "provide and regulate arterial, toll, and other roads . . . tunnels, and related facilities . . . "; to "provide and operate air, water, rail, and bus terminals, port facilities, and public transportation systems"; and to "license and regulate taxis, . . . operating in the unincorporated areas of the county."[16] Similarly, in Toronto, the Metropolitan Council is given virtual total authority over the "metropolitan road system" (including the power to designate which roads are of metropolitan significance; only sidewalk maintenance is exempted from potential metropolitan control) and over "metropolitan transportation" through a Toronto Transit Commission whose members are appointed by the Metropolitan Council.

Undoubtedly some such reallocation by function is appropriate, but we should be concerned lest we deceive ourselves into believing that transportation functions can be neatly divided into those primarily of neighborhood concern, those primarily of metropolitan concern, and those primarily of state concern. The earlier discussion of expressways should have made clear that there is a spectrum of concern about their location, their scale, their design—all the way from the impacted neighborhood, through the particular suburban corridor, to the greater metropolitan area, the state, and even the federal government. Even arterial streets, which may seem to be primarily of

---

[15]*Ibid.*, p. 7.

[16]The Charter of Metropolitan Dade County, Florida, May 21, 1957, Section 1.01 (A) (1, 2, 3).

metropolitan significance, can often act as the main commercial street of a particular municipality or area (Massachusetts Avenue through Cambridge and Arlington, Massachusetts, or the Grand Concourse in the Bronx). Moreover, particular aspects of arterial streets, such as traffic lights or stop signs, can have very local significance. One of the early community-organization training films coming out of Newark in the early 1960s showed the use of neighborhood feeling about the absence of a stop sign as the opening wedge for an Alinsky-type organizing effort. Transportation, in other words, is and should be as much of a "marble cake" as any other governmental function; nor can it be rationally allocated by neat subfunctional categories, each going in its entirety to one governmental unit or another. There is, properly, a mix of institutional responsibility that reflects a mix of legitimate interest in each subfunction.

This analysis leads to two conclusions. One is that in any major experiment in creating new entities for metropolitan governance, we should pay as much attention to creating legitimate neighborhood or community entities as we should to creating the overall metropolitan entity. This point is made with force in the CED Report: "In addition to an area-wide level, modern metropolitan government should contain a community-level government system comprised of 'community districts.' "[17] Each such community district would in many cases be composed of a subsection of the present central city. A single suburb, or a group of several, could form another such district. Such a two-level system has great potential in dealing with transportation problems. It should lead to more effective representation of neighborhood interests and provide a forum for the jockeying of those interests into what one could call a metropolitan viewpoint.

The second conclusion is that metropolitan transportation problems as they have been described in this paper require the establishment of these two levels of forums, with the possibility of creative (as well as disruptive) interaction between the two, more than they require a clear prior allocation of transportation subfunctions to the one level or the other. It is not subfunctions that should be assigned; rather, it is different kinds of power over each subfunction that should be allocated.

This paper does not set forth what allocations of power over transportation functions would be appropriate within the context of a broader metropolitan governmental reformation. The nature of some of the powers that might be allocated, however, will suggest the kinds of relationships that should be considered. As a start, the following powers may be listed:

(1) The power to veto, absolutely.
(2) The power to veto, subject to override.

[17] *Reshaping Government in Metropolitan Areas*, p. 20.

26 Metropolitanization and Public Services

(3) The power to delay for a specified period of time.
(4) The power to propose.
(5) The power to advise.
(6) The power to review and comment.
(7) The power to appeal (to the next higher step in the hierarchy, or to a different branch of government).
(8) The power to go forward only if a specified majority of a council or legislature has been achieved (simple majority, absolute majority, two-thirds, three-fourths, etc.).

This list sounds rather formalistic. But where such forms reflect political realities and political equities and the checks and balances that those realities and equities demand, then they have a key role to play in creating a forum within which the various interests and values can plan the game of politics.

Such a change in governance is a task in constitution making—"making" instead of "writing" because constitutions are made; they are not made up. They come out of the real problems felt by real groups in particular historical, geographical, and institutional settings. There will, therefore, be great variations from metropolitan area to metropolitan area.

Nevertheless, one can outline some general points of similarity that would be likely to emerge:

(1) A legislative council, which could be composed of the present (or reapportioned) state legislators, who would thus sit in two councils.[18]
(2) A single executive officer, elected by popular vote of the entire metropolitan area.
(3) A judicial branch to interpret metropolitan ordinances.
(4) Possibly, a regulatory branch (for instance, to hold the hearings on major transportation plans outlined earlier in this paper).
(5) Procedures for the creation and governance of community districts.
(6) A metropolitan bill of rights.
(7) The possibility of adding new "community districts" as a metropolitan area expands (an attempt to deal with the problem of the continual growth of metropolitan areas).
(8) An amendment process.
(9) A ratifying and convening process—a metropolitan constitutional convention? Convened by whom?

Some of the schoolboy slogans about the federal Constitution are worth recalling, as we think about a metropolitan constitution: checks and balances; majority rule with minority rights; taxation only with representation; the

---

[18] See a discussion of this idea, and a critique of Councils of Government in Frank I. Michelman and Terrance Sandalow, *Materials on Government in Urban Areas* (St. Paul, Minn.: West Publishing Co., 1970).

President as the spokesman for a national constituency; the integrity of an independent judiciary; the separation of powers.

Most of these principles have immediate application to our metropolitan problems, although it is worth noting that for transportation purposes I am more interested in checks and balances and in equity (protection of the transit-riding minority, for example) than in a separation of powers by function.

Perhaps these thoughts are either too frivolous or too formalistic. Perhaps we have a working metropolitan constitution already—although I would argue that across a broad spectrum of issues, not just transportation, such patterns of metropolitan governance as we have are mainly inefficient, inequitable, and illegitimate. My own hunch is that our metropolitan areas are ripe for the kind of politics that could push for changes along the lines suggested here. At least in some parts of the country, there is a growing feeling that central-city problems and suburban problems cannot be solved separately, that they are really all part of a common, metropolitan problem. In Boston, for example, a joint central city-suburban coalition revolving around transportation issues has emerged as a significant political force within the last two years. Perhaps metropolitan transportation politics are a sign of what may be coming in other areas, given effective political leadership to articulate the cause and intellectual constructs available for such leaders to adopt and adapt.

2 Education and Metropolitanism

DAVID L. KIRP AND DAVID K. COHEN*

The rationale for rearranging school government along metropolitan lines includes everything from management efficiency to racial integration. The policy of district consolidation for reasons of efficiency reaches back several decades, to the outset of the school-consolidation movement: 50 years ago the nation included more than 100,000 districts; today there are less than 25,000. Most of this consolidation has taken place in rural America, but there has been a good deal within metropolitan areas, and there are a few southern metropolitan areas that now have countywide school systems. There is, however, no sign of any major movement toward metropolitanism in education.

The reasons for this are not obscure. Metropolitan school government could spell the demise of existing local school agencies, and it would be likely to redistribute resources. But existing governments have their own vested interests, and the affluent are reluctant to surrender control of resources.

If community self-interest or myopia were the sole obstacle, however, the appropriate strategies would be fairly clear: inducements for suburbanites (the most likely objectors) to make metropolitanism fiscally attractive, legal prods for communities whose resistance was grounded largely on racial considerations, and federal support for regional planning. But this argument ignores a basic question: Is metropolitan government a sensible way to address the problems of urban public education?

*David L. Kirp is Acting Associate Professor, Graduate School of Public Policy, and Lecturer, School of Law, University of California, Berkeley. David K. Cohen is Director, Center for Educational Policy Research, Graduate School of Education, Harvard University, Cambridge, Massachusetts.

Several points bear on this query:

(1) Inequities in school revenues and costs within the metropolis.
(2) Inefficiency in school operations caused by the proliferation of school districts.
(3) School segregation.
(4) Rigidity and uniformity in the quality of education and poor matching of school programs with client demand.
(5) Lack of much parent involvement in, or control over, school decisions.

We shall pursue these five points in the rest of this essay.

The problems are highly visible in the metropolitan areas, but this is no guarantee that they can be alleviated by metropolitan government. Nor does the presence of problems in the metropolis mean they are absent elsewhere. Some problems may really be statewide, and if this is true, metropolitan government may be inappropriate. And even if the difficulties are uniquely metropolitan, that does not lead inevitably to a single broad and more inclusive governmental structure. Different metropolitan structures may be needed to address different educational problems in the same metropolitan area.

Traditional Arguments for Metropolitanism

The traditional arguments for metropolitism in education have turned on problems of school finance, race, and efficiency. Although none of these seems very convincing to us, the arguments have dominated discussion up to this point, and should be disposed of before turning to more recent concerns.

Fiscal Disparities

The discussion of metropolitan fiscal problems has been underway for more than a decade. Central cities are believed to have less money for schooling than their suburbs. This is attributed partly to the greater competition for tax revenues in cities, partly to the weaker city tax base, partly to higher school costs in cities, and partly to the presumption that children in the cities are more costly to educate than their suburban counterparts.

This widely accepted argument does not entirely square with the facts. Most central cities do not have less money for schooling than their suburbs: at worst, the cities are close to the average for local expenditures within their metropolitan areas. But if the poverty of city school districts is not reflected in an absolute disparity in revenues, it is evident in several other ways. The older northern city school systems have grown less affluent relative to the average suburb. It is necessary, however, to diversify our conception of the jurisdictions included in the designation "suburb." In reality, the central cities' fiscal problems represent only one manifestation of the fiscal burdens that afflict older northern urban school districts of any size or location. These districts experience higher-than-average competition for revenues, a tax base depressed by low residential valuation and industrial blight, and more poor children than most other metropolitan districts. (In Massachusetts, for example, the public schools in roughly a dozen such cities enroll more than nine out of every ten children from welfare families in the entire state.) Thus, the "crisis of the central cities" is not as neatly limited as the phrase implies; indeed, it is the crisis of all older industrial cities, whether "central" or not.

It is easy to imagine ways of correcting this situation. But, apart from creating a metropolitan tax district, none of these solutions would be uniquely metropolitan in character. One approach has been proposed by Governor Milliken of Michigan; its keystone is the establishment of both a uniform statewide tax rate and a fully equalized assessment. Districts would be taxed uniformly, and the results distributed in such a way as to take cost variations into account. Another approach, urged by John Coons, would revise state school-aid apportionment formulas so that an equal tax effort in any two districts would produce equal revenues, irrespective of local wealth.[1] The crucial difference between the Milliken plan and Coons's scheme is that under the latter full local autonomy could be retained in deciding the level of effort.

Creation of a metropolitan tax district would not be a helpful approach. It offers no fiscal advantage that a similar statewide plan would not, and it has the additional disadvantage of dealing with only part of a problem that is indisputably statewide. Even apart from such objections, a metropolitan tax district would probably be more difficult to arrange. It is inconsistent with the constitutional standard advocated by John Coons and his colleagues, and recently accepted by the California Supreme Court in *Serrano v. Priest*, which renders illegitimate any education revenue-raising scheme tied to local property wealth. For that reason, a court would be unlikely to take seriously a suit asserting the unconstitutionality of metropolitan—rather than statewide—taxing inequities. And if such an arrangement were proposed in the legislature of any urban state, it probably would not get far. The central cities must be able to gain enough additional support in smaller nonmetropolitan cities to make a statewide plan much more attractive as a vote-getting device.

Metropolitan government, then, seems to offer a poor approach to the alleviation of fiscal disparities among school districts. Although the metropolis will continue to be a focus for discussion of this problem, solutions will be statewide, if they are possible at all.

Efficiency

The traditional argument for school-district consolidation is that larger units permit quality educational programs at a lower cost than do smaller

[1] J. Coons, W. Clune, and S. Sugarman, *Private Wealth and Public Education* (Cambridge: Harvard University Press, 1970).

schools or districts. This argument has been such a standby in the struggles over school-district consolidation that it has become a part of American educational mythology.

This might not seem odd were it not for the ambiguity of research results. There have been literally scores of studies that have related the school characteristics in question to student achievement; and almost without exception they have found no consistent connection.[2] The size of schools and school districts, for example, has no regular association with student achievement: students in bigger schools or districts typically do no better than similarly situated students in small ones. Moreover, the things that larger districts are supposed to provide more of—experienced and well-trained teachers, better curricula, more differentiation among students by interest and ability, more libraries, better administrators, and more specialists—also have little relationship to achievement. Students in schools with more of these educational accoutrements seem to do no better than similarly situated students in schools with less of them. All of this was evident in hundreds of small studies carried out over the last 40 years; and it was confirmed in two massive surveys of American education completed in the last 10 years—Project TALENT, and the Equality of Educational Opportunity Survey. This research has shown rather persuasively that the things that most Americans believe distinguish good from bad schools do not transform bad students into good ones. School consolidation, apparently, will neither improve students' test performances nor the schools' efficiency in producing achievement.

Of course, students' test scores are not the sole criterion of school efficiency. Other school outcomes, such as aspirations or the ability to participate in public life, may be affected by the size of the schools that children attend or the sort of students in their classes. But the evidence on these points is also ambiguous. Students in smaller high schools, for example, seem to participate more, and to develop more self-reliance, than otherwise similar students in large schools.[3] This hardly augurs well for the advocates of large schools and districts. And the evidence on aspirations suggests that most of the differences lie among students within schools rather than between schools. That is, the variation in aspirations within American high schools is little different from the variation in aspirations among schools. Thus, redistribution of students among schools would have little effect on the distribution of aspirations.[4]

---

[2] The best review of research in this area is found in J. M. Stephens, *The Process of Schooling* (New York, Harper & Row, 1967).

[3] R. Barker and P. Bump, *Big School, Small School* (Stanford: Stanford University Press, 1964).

[4] These results arise from as yet unpublished analyses of data on high school students' aspirations, which derive from the project TALENT and Equality of Educational Opportunity surveys. The results are available from the authors.

The only convincing evidence for the efficiency of consolidation is that administrative costs seem to be somewhat lower in medium-sized districts (between 30,000 and 50,000 students) than in very small ones. This is of little real importance, however, since the annual per student cost of these differences amounts to less than $3 or $4 per year, and the diseconomies of scale increase fairly steeply as district size rises beyond 50,000.[5] If the only form of consolidation that might promote greater efficiency were the creation of several medium to large districts within each metropolis, the savings would be too small to override the political resistance to this undermining of local autonomy.

These research results undercut most of the traditional justifications for creating larger schools and districts. Other arguments, of course, remain. Neither popular taste in education nor the desires of school professionals are likely to be affected much by research. Larger schools and districts are attractive to professionals because they mean newer buildings, more specialization, and more pleasing facilities: libraries, language labs, offices, etc. Since the public accepts these as valid indices of educational quality, the old arguments may still produce some support for consolidation.

Integration

This subject has been the featured item in almost every recent discussion of schools and metropolitanism. Neither the rationale nor the remedies have changed much in the last decade: The educational benefits to Negro children and the impact on black and white attitudes are the main arguments used to justify integration. The present, or prospective, black-majority character of many central-city school systems is the reason for seeking a remedy that will tie cities and suburbs together.

Perhaps the central issue here is whether the rationale is correct: Would integration have an important effect on achievement or racial attitudes? Even if the answer to this query were unambiguously affirmative, other issues would arise. What form should the new arrangements take: Unified metropolitan districts? Interdistrict cooperation? The creation of several large, but submetropolitan districts? Would a metropolitan school authority, or a state agency, be the best device to further integration?

Evidence on the effects of integration is far from unambiguous. Most research suggests that the schools' racial composition has no independent impact on students' achievement, but that their social-class composition does have some effect.[6] Disadvantaged students in middle-class schools seem to do

[5] The data on district size and administrative costs are contained in a forthcoming study of school decentralization, "Parents, Power, and the Schools," which is available from the authors.

[6] For a summary of the evidence on this entire issue and a discussion of its implications for law and policy, see D. Cohen, "Defining Racial Equality in Education," *UCLA Law Review*, vol. 16, no. 2 (February 1969).

somewhat better than similar students in uniformly working-class or poor schools. If true, this means that school integration within cities (where blacks would typically be integrated with whites from adjoining blue-collar neighborhoods) would be less desirable than city-suburban integration (where blacks would be more likely to attend school with middle-class or upper-middle-class white students).

However, it is by no means clear that this finding is true. Black students in white schools, or lower-class students in middle-class schools, might be there because of their greater talent and/or their parental motivation, which would account for their higher scores. Since information on students' earlier academic or family history is rarely gathered, it is impossible to determine whether the effect of class composition is due to such selection.

Evidence from deliberate desegregation efforts is also ambiguous. In some cases in which achievement gains appear, students have almost surely been selected by IQ; in other instances (whether there has been selection or not) there seem to have been no gains. In the only case in which students were randomly chosen (to eliminate the selection problem), small gains did appear after integration in some instances. Compared to the magnitude of the black-white achievement disparity, however, these gains were quite small. Only one study has carried out such a check (in a very small California city); it was found that if one took first-grade IQ into account this very sharply reduced, but did not entirely eliminate, the social-class-composition effect on later IQ and achievement tests.

The research results on racial attitudes are more consistent and encouraging. Blacks and whites who enjoy equal status tend to exhibit less prejudice. If they had such experiences as children, they are somewhat more likely to have interracial contact as adults, and they are a good deal more likely to hold more liberal attitudes on public issues related to race. These findings accord with common sense and general observation, and they suggest that if school integration were properly managed, it might have some effect on American racial attitudes and behavior.

But however affirmative the evidence, it does not lead directly to integrated metropolitan school districts. There is little experience with school integration within the metropolitan context, and only modest thought has been given to the issues. The experience encompasses three or four interdistrict busing programs and involves only a few thousand children. Although programs of this nature may serve to demonstrate the feasibility of interdistrict cooperation, they have serious drawbacks. One-way traffic from the inner city, after all, is hardly a promising model to apply to the entire metropolis.

Other approaches have been suggested. The U.S. Commission on Civil Rights, for example, recommended federal fiscal incentives for the construction of educational parks through metropolitan areas.[7] Presumably such an arrangement would require the creation of unified metropolitan education authorities. Although possible, this plan would almost surely be dependent on action by the state legislatures, something that legislators from many suburban districts would resist.

Another alternative would be the creation of several large districts composing both core-city and suburban neighborhoods. Although in several of the largest cities such a scheme would be unfeasible without helicopters or new rapid-transit systems, there are many more places where it would be technologically possible. In addition, such new jurisdictions, were they to enroll somewhere between 30,000 and 50,000 students, would approach administratively what seems to be the most efficient size. But again this approach would almost always require legislative action, and it is no easier to imagine Shaker Heights or Winnetka merging individually with parts of Cleveland or Chicago than it is to imagine an entire city merging with all its suburbs. Racial considerations aside, it seems politically fantastic. Would municipal administrations voluntarily relinquish the money, jobs, and power inherent in a big-city school system any more readily than would affluent suburban systems? In fact, city-suburban mergers will probably become less likely as the central cities move increasingly toward black self-government! These new administrations will badly need the money and jobs on which political organizations are based. As long as public education in the metropolitan areas retains even a tenuous fiscal viability, it is hard to conceive of the system of state or federal incentives that could produce wholesale movement in the direction of metropolitan school integration.

Other problems also exist. The most serious such problem is that pressures for other changes have recently gathered force, especially in the older and larger metropolitan areas. There has been a rapid increase in demands for greater community control in the black ghettos; it is possible to theorize that integration is consonant with community control, but it seems unlikely to be true in practice.

Is there, then, any prospect of a substantial movement toward school integration within the metropolis? If the vehicle is voluntary local action, the answer is no. If we turn to the federal government for a moment, the picture is not vastly more hopeful. In the near future, at least, executive action and new legislation will at best offer no more than support for local initiative toward integration, and it may offer some discouragement. Although both local action and federal support would be worthwhile, there is no reason to believe that the results would affect many children. Experience with the

---

[7] *Racial Isolation in the Public Schools* (Washington, D.C.: U.S. Civil Rights Commission, 1967).

existing metropolitan busing programs—in Hartford and Boston—suggests that they do not lead to more integration. Rather, once they achieve a modest percentage of blacks in the receiving schools, the situation seems to stabilize.

Litigation is the remaining approach to metropolitan integration, and it may well prove the most promising. Courts have shown increased willingness to identify northern school segregation as *de jure*—based, that is, upon pupil and teacher assignment, school building patterns, and discriminatory housing—and to order integration *within* city boundaries. In several suits, litigants have asserted that the discrimination is not limited to school districts but is a metropolitan phenomenon; they have demonstrated areawide housing, job, and school boundary determinations that serve to isolate black and white children. At least two federal courts, in Detroit and Indianapolis, have accepted these arguments, and ordered the submission of metropolitan integration plans; other such suits are currently before the courts. While such litigation may continue to force metropolitan governance in particular northern communities, the process is likely to be an extended one, and ultimate resolution of the underlying constitutional question by the Supreme Court seems several years away.

Thus, even the most optimistic assessment of the prospects for school integration within the metropolitan areas indicates that only modest progress can be expected in the next decade. Furthermore, the most optimistic assessment may not be warranted. Conventional approaches to integration are at cross-purposes with other forces in the Negro community; and these other forces seem likely to dissipate the thrust toward school integration.

In summary, then, the traditional arguments for metropolitanism in education are generally unconvincing. Larger schools and larger districts might satisfy educators, but there is no evidence that they will either help students or appreciably improve the effectiveness of schooling. The fiscal disparities that plague districts within the metropolis need correction, but remedies should come at the state, not the metropolitan, level. Remedies for segregation are sadly needed, but there is little impetus and less likelihood that more will soon materialize.

There is, however, a further difficulty with any discussion of metropolitanism in relation to such problems as fiscal inequity, segregation, and efficiency. To argue that these problems require metropolitan government is to assume that the existing structure for decision making about schools is basically satisfactory and that the real problems lie in the allocation of students or resources. There is growing evidence that this assumption is incorrect. One example is the persistent pressure for community control in black neighborhoods, and another is the growing use of nonpublic schools by the urban middle and upper-middle class.

Both developments are signs of a deeper problem—the public school system's inability to respond to variations in client preferences. The supply of education in urban America is relatively uniform, but the demand is for diversifying it. There has been increasing concern with the delivery and decision mechanisms in education, which must be accompanied by a decline of interest in problems of equity and efficiency. What is more, these developments are uniquely urban, for only in the metropolitan areas do diverse client preferences exist.

Some Newer Concerns about Metropolitan Schools

This section of the essay takes up the questions of service delivery and governance for metropolitan area schools. Although these issues lie outside the traditional discussions of metropolitanism in education, it is no longer sensible to discuss equity and efficiency in the provision of public services without also considering the underlying delivery and decision-making arrangements.

Diversity

School reform in America has not always been distinguished by a concern for educational variety. Schoolmen are fond of calling for more individualized instruction, but the development of public education during this century has been largely the story of its standardization. This has resulted partly from the growth of the educational professions, but it has also had something to do with the schools' status as a public monopoly. Effective management has been equated with equality in the provision of resources, and sameness in the character of school offerings. This has been based both on a commitment to egalitarian values and on the view that the schools provide the principal vehicle for transmitting a common culture. The first has seemed to demand the provision of equal amounts of public education in equal quality, and the second has seemed to prohibit variations in educational taste (whether these have had an ethnic or class basis) from determining the character of school offerings.

In the last few years all this has come under increasing attack. The demand for more open, spontaneous, and creative schooling has grown (witness the hundreds of nonpublic schools now in existence), and the evidence that traditional school resources do not affect achievement has served only to reinforce this development. A number of critics argue that the main problem with public education lies in the fundamental assumption that it should yield uniform results and that it should ignore differences in educational preferences. As things now stand, families can provide a different education for their children only by changing residence or by enrolling in private schools. These possibilities are hardly adequate to ensure a match between client taste and school offering. The fiscal constraints against opting out of the public schools are severe, and the lack of direct public support for any alternatives limits sponsorship either to large-scale enterprises (like the Catholic Church) or to very wealthy small ones.

The remedies that have been advanced are tailored to this diagnosis: the cure is seen to lie in providing public support for competing institutions. It is of little practical comfort to parents that they have the right at present to send their children to whatever school they wish when they may have no money with which to exercise that option. Among the arrangements most commonly mentioned are funding for private schools (parochial schools, boarding schools, community schools, free schools, etc.), eligibility of new schools for state financial assistance; or the provision of direct financial support to parents—through tuition vouchers that could be used at any schools they chose.

The implications of such proposals are enormous. Neither direct funding of schools nor direct funding of students would remove the state from the business of public education. It would still provide financial support; it would regulate schools to ensure that they complied with constitutional requirements (for example, that they did not discriminate on the basis of race and that they provided a minimally acceptable level of education). Further, it would continue to maintain the existing public schools for those families who chose to patronize them. But the state would no longer offer a single form of schooling as the only possibility for parents too poor to send their offspring to a private school.

Such schemes (a test of tuition vouchers may soon be financed by the Office of Economic Opportunity[8]) are often thought to conflict with the need to redress fiscal inequity or race and class isolation. It has been argued, however, that diversification could be structured to increase fiscal equity. Equal dollars could be provided for each child, and the state could prohibit parents from supplementing that allotment. In addition, the state could provide more money for poor children—or for schools that accepted substantial numbers of poor children—thus giving them (theoretically) a better bargaining power. (This second approach might also reduce race and class isolation by inducing schools to take more poor, and therefore black, children than they might ordinarily.) In the case of fiscal equity, then, there is no necessary conflict. Were some statewide, full equalization plan adopted, it would not necessarily interfere with either direct aid to alternative schools or direct tax subsidies to families.

The problem of segregation is less easy to dispose of. Tuition vouchers would presumably permit parents to choose whatever school they liked. Black families in mainly black districts who desired integrated schools for their children would not be restrained by geographic school attendance boundaries, any more than would suburban white parents who desired

[8]The planning has been under way at the Center for the Study of Public Policy in Cambridge, Mass.; a report on the program, *Education Vouchers*, is now available.

Montessori schools for their offspring. This, of course, is the great appeal of tuition vouchers; they could permit communities of limited liability to spring up around schools, thereby allowing expression of diverse tastes. But therein also lies the great difficulty of such schemes; for the individual freedom to choose schools could also be the freedom to restrict the attendance of others. Black parents might want to enroll their children in white schools, but the whites might have other ideas. An unrestricted voucher scheme could harden patterns of segregation even more than at present, and that would be a disaster.

There are devices that could avoid this problem, at least in theory. If racial discrimination in admission were prohibited, and if fiscal incentives (of the Title I, Elementary and Secondary Education Act, variety) were offered for the enrollment of poor children, many of the problems would disappear. If, in addition, a considerable proportion of applicants at every school were admitted by lottery, there would probably be more school integration than now exists.[9] The difficulty, however, is not in conceiving of schemes whereby segregation could be avoided; rather, it is in imagining how such schemes might be implemented. It is hard to see how even such relatively progressive states as New York and Michigan would approve legislation of this sort, when they have so recently been moving in the other direction.

Apart from this problem, however, our question is whether direct tax subsidies to schools or parents would be suitable at the metropolitan level.

Diversity becomes an issue only in a community large enough to include many families with quite different educational interests. It would make no sense to adopt a voucher plan in a small community in which parents shared similar perceptions of educational needs. But in the metropolitan regions such uniformity does not prevail; the differences and conflicts are all too apparent. Only in a community of that size would diversification accomplish anything new.

A metropolitan educational authority might well manage the sort of system we have described. Such an authority, unlike a metropolitan arrangement structured to promote integrated schools, should probably not operate schools in the metropolitan area. Instead it would function only as a certifying and disbursing agency: it might evaluate schools, check teacher qualifications, and assess the accuracy of schools' self-descriptions. It could either distribute checks directly to those schools that satisfied the authority's standards or it could distribute tuition vouchers to the parents who could then cash them at approved schools.[10]

---

[9] These ideas, and others, have been advanced in the study *Education Vouchers* referred to in n. 8.

[10] It would, in fact, be necessary to devise ways to ensure that such a step did not increase bureaucratic control over education. The creation of a metropolitan educational authority with its own priorities and standards might represent simply another layer of

Other arrangements are imaginable, such as placing the governance of such a system in the hands of existing school districts, or of the state. The first of these—district-controlled systems—would minimize the need to tinker with the structure of educational governance. But this might be a disadvantage, because it would leave the responsibility for change with the very parties who most vociferously resist the idea. It seems unlikely that the agency that already operates the public schools, and that would continue to operate those schools for the children who still elected to go to them, would effectively support educational diversity when it meant encouraging children and parents to opt out of attending those schools.

State management represents the other alternative, but it might be inappropriate. Uniformity of education offerings is not solely a metropolitan problem, but the diversity of taste in towns, small cities, and rural areas is much less than in the metropolis. Theoretically, a state-run system would permit greater educational diversity by enabling students to attend school outside of the area (city or metropolitan) in which they happened to live. The chief disadvantage is that it might be impossible for a state agency, quite removed from the actual provision of education services, to administer intelligently a voucher or direct school-funding scheme.

On balance, a metropolitan authority seems most sensible. Although its existence would require special state enabling legislation to set certain standards (and presumably to create similar authorities in nonmetropolitan counties as well), a metropolitan agency would be large enough to comprise sufficient resources for diversification, and yet not so distant as to be unable to regulate, disseminate information, experiment, and evaluate. Although the diversification of schooling is not a uniquely metropolitan matter, it is something that a metropolitan educational authority would be uniquely suited to manage.

There are, of course, quite possibly fatal political difficulties with such an arrangement. Local school districts are unlikely to surrender power willingly, and to the extent that a metropolitan authority would blur city-suburban distinctions, opposition might be swelled by suburban parents eager to preserve local autonomy. There is certainly no evidence of real movement in this direction at any level of government.

Community Control

Schemes for diversification would assuredly benefit Negro families; yet there has been little black support for tuition vouchers or direct funding for

---

bureaucracy. But such an authority could be structured to reduce government's power over the provision of educational services, by taking the authority out of the business of operating any but a small number of experimental schools, by leaving school operation to the existing public districts, and/or by limiting the kinds of regulations that the authority could impose.

alternative schools. Rather, attention has been fixed on the public schools that serve Negro communities, and demands have centered on efforts to make those schools prove responsive.

Proposals for community control and individual choice of school proceed on inconsistent assumptions about the nature of the school community. Advocates of individual choice presume that parents may want different kinds of education for their children; that in the absence of present constraints the choice of school would not necessarily be determined by where a family lives. Thus, one ghetto family might choose to send its children to an all-black community school, but another might opt for an integrated suburban school. Tax subsidies to schools or parents would leave that choice to the family. Advocates of community control, on the other hand, regard the geographic-ethnic community as the proper focus for school attendance and decision making. They seek to give the poor and black communities the power to run neighborhood schools in a manner that they feel will more adequately address community needs.

The demand for community control is frequently couched in educational terms. Such control, it is argued, will afford the child a sense of mastery over his environment, and this will enable him to succeed; teachers in such schools will be obliged to respond to the particular needs of the children in their charge; parents will be more involved, and children will therefore perform better. Educational evidence to support these propositions is lacking, but advocates argue that communities could run their own schools no less successfully than the existing bureaucracy.

The demand for community control is, of course, in good measure political. The result would be real power: the right to determine how education dollars are spent and who gets what jobs. But whether viewed as an educational or political effort, community-run schools address different concerns from those that motivate other reformers to call for an end to racial isolation, or for increased educational diversity, or even for more equitable distribution of educational resources. To be sure, one can imagine heterogeneous school districts. There is no intrinsic reason, for example, why Bronx-Westchester or Queens-Nassau districts could not have been created when the New York State Legislature was considering alternative ways to decentralize New York City's schools. That such arrangements were not even proposed, let alone considered, reveals the political barriers to such a notion. Moreover, the ideology of control depends on the assumption of homogeneity and likemindedness.

It is not clear, however, whether this inconsistency between direct tax subsidies to parents and demands for community control is inevitable. Public education has been the focus of attention in the ghettos for more than a decade, and it is very nearly the sole source of schooling available there. Given these circumstances, and the relative inability of blacks to afford alternatives, it is not surprising that efforts have focused on diverting public resources to community use. But a system of tax support for parents, or alternative schools, would serve most of the educational and political ends sought by advocates of community control and yet provide more choice for ghetto residents. What is more, this scheme would offer a way around some of the more obvious political problems of decentralization. Chief among these is the possibility of a fiscal crisis, which could arise if black ghettos tried to support themselves from the existing tax base. Of course, legislatures alone could correct this, but black communities would find no new allies on this point beyond those already clamoring for revision of state-aid formulas.

In any event, decentralization is not really relevant to the metropolis, at least not at this point in time. It would offer the cities much conflict, and the suburbs could gain nothing from it.

Conclusion

In general, then, the prospects for relieving the educational problems of the metropolis through more comprehensive government are bleak. A remedy for the underlying fiscal problems is available only at the state level; larger districts promise nothing; there is no likelihood of anything but minimal progress toward integration via metropolitanism; and city school decentralization is really irrelevant to most jurisdictions in metropolitan areas. The only governmental revision that would be both relevant to metropolitan educational problems and also appropriate at the metropolitan level is the most radical one: a system of direct tax subsidies or direct aid to nonpublic schools, which would transfer school decisions much closer to the family and to the individual school.

A system of tax subsidies whereby information, disbursement, quality control, and evaluation lie at the metropolitan level, and everything else is left with the schools or districts, would probably be an improvement on existing arrangements. Such a system could resolve many of the problems that presently plague schools and parents in metropolitan areas, such as unresponsiveness, rigidity, a poor match of demand and supply, and the burden of accumulated routine and bureaucracy. A properly implemented system would not increase segregation, and might reduce it.

Such arrangements, however, are unlikely soon to see the light of day. The demand is just beginning to become substantial, the church-state issues may not be fully settled for some time, and most of the established school authorities are opposed. The perception of educational problems within the metropolis is still too uneven for a workable consensus to be built around this particular scheme or other program for change. Things may have to become appreciably worse throughout the metropolitan regions—especially in their more affluent sectors—before public education has much chance of being put on a new footing.

3 A New Look at the Health Issue

LEONARD J. DUHL*

The present symptoms of our society, its current turbulence and counterturbulence, are but signals to indicate that our "ability to command events that affect our lives" is markedly impaired. Thus, the current search toward metropolitanism is only one part of many parallel, but similar, attempts to deal with the "transactional"[1] issues of interrelationship among institutions, values, social systems, and geographic areas. It is within this context that metropolitan development is emerging. Therefore, this paper will outline in general terms the factors leading toward metropolitanism. Next, a discussion of the health field to support and expand the thesis will follow.

The processes currently taking place in society, which resulted from an expanding technology, have created new aspirations, new values, and demands for great change in our institutions. This shift toward more concern with values and human needs than with technological development is a critical one for understanding our future development.

The basic psychological nature of man is conservative. He resents change (except in limited areas), hates bigness, and dislikes external control over his personal behavior when it interferes with his self-perceived needs. He is also, especially in these times of crises and change, ideologically conservative. Despite, or perhaps as a result of, confrontations with his ideology, his usual

*Leonard J. Duhl, M.D., is Professor of Urban Social Policy and of Public Health, College of Environmental Design, Department of City and Regional Planning, University of California, Berkeley.

[1] "Transactional" issues relate to the many systems questions of interrelationships: communication, language, interpersonal relationships, and the rules of intergroup interaction, plus the many tools we use to aid interaction, among which are money, symbols, and laws.

response is to opt for unchangeability, the familiar, and the political status quo.

Pragmatic, immediate issues are central to man's concerns. Pure practicality—on such issues as economic return from investment and its impact on his way of life—lead to self-directed demands for new ways to cope with his day-by-day "bread and butter" issues. The pressure to meet these "self-directed" issues and to establish personal economic security, while maintaining ideological conservatism, has led to pragmatic changes in a variety of areas, such as business, health, transportation, child care, social welfare, and communications. Man may be *ideologically conservative* but he is *pragmatically radical*. In a pragmatic area such as health, where the major crises affect every interested party from the providers to the suppliers and consumers, there are many attempts to search for new solutions to problems. In these attempts, new patterns of governance are developing and will continue to develop.

The current fragmented sources of federal funding are being modified gradually toward becoming special-interest block grants; increasingly the only source of funding will be government, federal and state. But it is already clear that most of these funds will be anticity and directed through categorical channels of the state government.

On the local level, homogeneous groups will demand more and more autonomy and control over their "turf." Many will be groups new to political power. These groups, moreover, will be antifederal and antistate. They will take up their sights and concerns within the city. It is in the city that man will put most of his effort to control those things that affect his life. But the turmoil will lead to very little resolution. Indeed, many of the noncity agencies, such as the state and federal government, will be less and less willing to deal directly with the city because of their fear of minority control. Thus minority-controlled cities will become isolated.

By default, then, the most important area of governmental activity that will emerge will be one that neither is involved in the day-by-day actions of the cities nor is part of state and federal bigness: it will be a metropolitan one. Metropolitan government will not emerge de facto, nor will it emerge from an ideological base. Out of pragmatic needs in a variety of fields, such as health, will come a collection of special-interest groups, authorities, and public/private authorities that will ultimately become forms of *metropolitan governance*.

The function of government in the national arena will increasingly be the provisions of funds. Concomitant with this will be a withdrawal from direct-action programs and the gradual development of performance specifications against which all lower-level governances should develop their activities.

The more one moves from local toward metropolitan government, the more important will become the process of holding together "the family of separate interest groups." The process of communications and linkages—the infrastructure—between the various subgroups will become the focus of primary concern.

The word "infrastructure" has heretofore been limited to those formal structures such as urban-area highways and communication systems that hold the parts together and mediate the transactions of the separate institutions. In the metropolitan area, the prime infrastructure will be the less formal structures holding the pieces together that involve the psycho-social and political interrelationships between people and groups.

New patterns of governance will arise on all levels, from the neighborhood to the city, metropolitan, state, regional (multistate), and the federal. The problem of linkage between these different levels will be as important as the "interfamilial" member relationships at any particular level. It will be metropolitan governance that will play the roles of broker, communication linker, and network manager among these various levels.

With the concern for local control, new patterns of power will emerge. Rather than power being delegated from a centralized authority to the periphery, it will be assumed increasingly by the autonomous local groups and then delegated to the city, which in turn will delegate it to increasingly higher levels of government. From the experience of local groups will be created the standards for each local community, and from these standards will come new performance specifications.

Problems of Health

It is now useful to turn to the problems of health. My basic thesis is that in each of many separate pragmatic and specialized areas of concern new processes of governance are being worked out that are unique to the needs of that area. In the field of health, for example, the crisis that currently exists is leading to a series of processes that are playing themselves out as new forms of organization and governance emerge.

It is my proposition that the phenomenon taking place in health is being replicated in other areas. Metropolitan government will not come from the new governances in health. However, as many of the special-interest, pragmatic developments begin to find new forms of governance on all levels, an ad hoc, loosely structured metropolitan government will begin to emerge. Health offers just one example of how metropolitan government will not be built out of ideological concerns, but rather of how new, informal, special-interest governance will prevail. As has been said already, we are ideologically conservative and pragmatically radical.

There are many problems faced by the health field. These range from questions of equity and quality of care to fragmentation, disorganization, and the high cost of services. A wide variety of people are involved in "the game of health," but they act as if they were playing separate games disjointed from one another. Insurance companies, hospitals, doctors, real estate operators, the pharmaceutical industry, and a host of other groups are involved in the current medical nonsystem. Each has defined his turf. Some are local and preoccupied only with the immediate needs of the area they are in—for example, hospitals and the private physician; some have national scope; still others international. Each goes his own way despite the fact that there have been attempts to rationalize the planning process. Still others plan comprehensively on a statewide level but are disconnected from the actual implementation of programs.

It is clear that there has been no careful analysis of what functions are most appropriate to what level of government. Which functions in health are primarily the concern of the local area? Which are citywide? Which are metropolitan, and which are state or regional? Furthermore, since the issues have not been clarified, there now exists a separation between planning and action that has prevented the implementation of even the wisest of plans. Health plans are seen sitting on the shelves of innumerable health planning boards in a manner familiar to the city planners' efforts of the 1940s and 1950s. Our purpose here is to discuss problems that exist on various levels of the current delivery of care, and to see how alternative models of governance and control will emerge to cope with them.

General practitioners, internists, and pediatricians are concerned with providing comprehensive general medicine. The specialist, on the other hand, is preoccupied with a particular fragment of the medical-care business. However, both the specialist and the practitioner have primary concern with giving the best possible care to those patients who come to them. They are less concerned about the people they do not reach than those that they do. The practitioner is usually so overwhelmed by his current practice that he is hard-pressed to consider the unreachable. He complains that he has too much paper work, that he is overwhelmed by the complexities of payment schedules, and that he is required to meet performance standards he has not played a part in setting. He thus feels that he is held responsible for the problems of the health care system and its inequities, although he believes that his primary concern is the delivery of the best possible care to those patients who come to him. His is a concern with local needs and a rejection of the issues of centralized power and the infrastructure.

Practitioners usually go where the population is. Moreover, the more doctors there are the more chances there are for new doctors to enter. Thus, those areas that are well-staffed have the greatest possibility of increasing their supply; those that are poorly staffed have little attractiveness to the physician. The private practitioner usually draws from a geographic area within his immediate radius, and the hospital similarly draws on its immediate, though wider, geographic radius. However, specialist practitioners and hospitals very frequently draw from much larger areas because of unique services they make available. Hospitals can be separated into those serving immediate neighborhoods and those serving a larger region, depending upon their level of sophistication, specialization, and the shortness of supply of personnel.

In addition to the hospitals there are a variety of other medical facilities ranging from nursing homes to extended-care facilities and out-patient clinics. These facilities respond in similar manners. Where no other facilities exist, emergency rooms open to all comers tend to act like magnets, drawing from areas far beyond the neighborhood.

In other words, most medical practice is focused on giving care to those people who enter the private system. Those who cannot pay—either through public or private means—are kept out, consciously or unconsciously, by economic factors.

New models are beginning to emerge, however. These models, unlike the old ones, are based upon population groups. Whether the population group is an industry, a labor union, membership in a group plan, or even a group practice, the responsibility to meet all needs of the population group leads to an immediate shift in the type of care given. Concern develops for preventive services as well as treatment.

When such responsibility for a population is superimposed on prepaid capitation health insurance,[2] the problem of fixed costs arises. The physicians and health personnel find that they have to operate within a budget, and they are forced to allocate their resources. Shifts take place in their priorities; expenditures for high technology and equipment, for example, become less important than immediate care. There is increased willingness to try subprofessional and ancillary manpower to improve efficiency. New patterns of ambulatory (nonhospital) care are evolving. These new procedures are emerging in the group practice field, in the neighborhood health centers for the poor, and in new extensions of hospital practice. In all cases there is clear evidence that the quality of care does not suffer when responsibility is increased to encompass a total population.

The preceding brief description does not include the influence of medical schools or the work of health departments or of numerous other health agencies. It does raise questions currently being faced by health planning boards that operate on the regional and state level. Only when responsibility to a population becomes a central issue do concerns such as metropolitan

---

[2] With "capitation," a single fee covers total medical care in contrast to "fee for service," whereby particular services are charged for at differing rates.

governance have any relevance to the participants. Responsibility for a population forces a concern with the infrastructure.

None of the existing health planning boards are capable of implementing their plans, primarily because they have no power to demand compliance. In certain areas the planning boards that control funds (for example, the Hill-Burton Hospital Facilities Boards),[3] have managed some allocation of construction resources on the basis of the needs of a metropolitan area. Where money is tight, especially, the Hill-Burton regional authorities have power; for most other planning boards, their lack of involvement in the action takes away their ability to influence the delivery of care.

The current health system, which is really a nonsystem (fragmented, unacceptable to consumers, and with little quality control or equity), cannot be held together by planning boards. Regional and metropolitan planning boards are supposed to integrate and coordinate health care activities, but they do not have the power, and no one really wants them to have it.

There is a further important issue involved in the medical nonsystem. This is the radical split in values within the system, very similar to the broader value split in the society at large. On one hand, the effects of science, technology, and the burgeoning of research in the 1950s and 1960s led to a vast increase of power for academic medicine in the universities. From this has come a continued push in the 1970s toward academic, scientific, and technological health developments. Indeed, much of the vast increase of health funding has gone toward implementing and developing the technological side of medical practice.

On the other hand, there are those primarily preoccupied with care and the needs of the clients. These clients, whether rich or poor, central city or suburb, rural migrant, minority group, or even sophisticated middle class, are having difficulty getting health care. As a result of fragmentation, the costs are going up. And the problem is exaggerated by the lack of rationalization. More important, the preoccupation with technology has created a vast chasm between the patient and the system. To the patient, the personal touch is more important than further scientific technological advance. Indeed, most of the demand for medical care is for *primary care*, in response to the patient's first complaint, "I hurt." However, most of what is available is *secondary* care, concerned with severe medical illnesses and disorders, and with it, high technology and science. There is a split between the humanistic personal health, general practice model—which is related to the more general values of society—and the scientific model of the current "power elite" in the medical profession, which mainly represents the status quo.

---

[3]Note that originally the federal Hill-Burton Act established primarily a state-directed program dealing with individual localities. Only when there was a need to negotiate over questions of "turf" and resources did it move toward a metropolitan orientation. It is my prediction that this will occur increasingly in all health areas.

A New Organization of Health Services

The federal mechanisms for care are disjointed and fragmented. Covering the broad range of needs at the local level is not possible because resources are so often tied to tightly defined purposes that do not include total population groups or community needs. Indeed, mechanisms for redirecting available money into broad coverage of services are lacking.

Most of the energy in this medical nonsystem, as in so many other areas, seems to be directed toward nonclient, yet presumably relevant, activities such as internal agency concerns, relations with other agencies, traffic flow, managing of money, and measures of efficiency. Only a small proportion of total resources (governmental and private) actually goes into direct services.[4] Much energy is spent in working out how that small proportion of activities directed to clients will be paid for and managed; much less is spent in finding ways to increase the percentage of resources channeled into real rather than support services.

The people most capable of organization for both medical and other care are often the ones most capable of obtaining funds and developing new programs, and they thus take a larger and larger share of the funds and the responsibility. The organizations with experience are the ones that are expanding, while those communities most in need of help have difficulty in obtaining money. The rich get richer and the poor get relatively poorer. New types of developments are extremely difficult to get started. Their proponents have neither the ability to deal with issues on an appropriate scale nor the funds to do so.

It is still unusual for all the people in health and related fields to come together for common decision making. When they do, the participants rarely act together; most often they fight. Indeed, there is no process of governance that permits the integration and working together collegially of the various groups. A "family" integrating mechanism does not exist for any problem solving.

It is now worth turning to one alternative answer to the problems. Clearly, there is need within a metropolitan region for the delivery of care on a more rational basis. Clearly, too, pluralistic alternative plans for giving care must be open to all community groups. The models of neighborhood ambulatory health centers that have emerged suggest not a pattern of government medicine but new forms of group practice, small enough in scale to maintain the personal level of care that patients desire and yet part of a larger network

---

[4] Over and over, in health, child care, education, and other human services, program directors turn outward in order to survive, to find funds, and to protect their turf, removing themselves from the prime responsibilities of the organization—the needs of the clients.

50 Metropolitanization and Public Services

involving hospitals and medical schools. These connections are required so that when complex problems arise in *primary care*, there are resources available.

This concept of care leads to a search for some form of central linking organization—perhaps a public-private authority[5]—on a variety of levels. "Linking" is used rather than "coordinating" because coordination implies that one is forcing separate programs to function as one, and that is not the objective. Rather, the need is for programs to be able to link when their objectives intersect but to exist otherwise as separate entities. It is the difference between a confederation of states and the United States. Such a central linking organization should include representatives of all programs, regulatory agencies, and consumers.

Some form of this organization should exist in the local neighborhood, and also in the citywide and the metropolitan regional levels. These various levels of organization should be able to provide for the provision of planning assistance to member programs; the assessment of communitywide needs and the creation or stimulation of new programs where deficiencies exist; the support of new program development; the provision of central services (for example, the training and supervision of professionals and paraprofessionals and the administration of such matters as personnel, budgeting, and bookkeeping for individual groups); the provision of specialist services; the interpretation of the system to the consumers; and negotiation with state and federal regulators.

A major challenge is to get the organization to come together and to design itself on its own initiative and its own terms. The potential for this happening exists in the commonly felt needs of the constituent actors, but the problem of providing a "catalyst" is a difficult one. Yet a solution to this problem is highly critical; a coordination organization that is imposed from outside is bound to fail. The organization must come together on the basis of its members' commonly felt needs, and these members must not feel that they are losing autonomy by participating.

The authority must receive its power, and thus its viability, through enabling actions on the part of those outside agents who now control resources and regulatory powers. The difficulties of catalyzing such a linking organization and of devising ways to judge its viability and the effectiveness of its use of resources are enormous.

Multiple-Level "Authorities"

In an immediate neighborhood, public-private "authorities" could be developed in which all bodies concerned with any problems, such as health care

---

[5] Or "public utility corporation."

delivery, could participate. In the first stage of development there should be a voluntary "forum" in which all the participants play a role in discussing and dealing with the issues involved. As the "forum"[6] begins to share ideas about problems, it could gradually incorporate itself into a locally governed public-private body to control the development of that issue. This local "authority" would be a new form of "congress," containing all participants in the system, including the consumer, and backed up by a staff. In the health field, the participants would include everybody in the medical network from the insurers to the pharmacists, other practitioners, and community representatives.

Each local area would participate by selecting representatives to send to a *citywide area "authority."* In no instance should the "forum" or "authority" take over the responsibility for running individual group programs or local activities. It should, however, reserve the right to develop new programs where needed by subcommunities and the community at large, and it should encourage others to increase their activities. Ideally these citywide and neighborhood "authorities" would have funds allocated to them for their activities.

But neighborhood or citywide "authorities" cannot deal with those problems that are *metropolitan* or areawide in scale, so yet another level would be required. Allocation of funds that will be increasingly available from federal and state sources could either go through, or be forced to "touch base" at, regional or areawide "authorities."[7]

This multiple level of governance by different levels of "authorities" has the major advantage that it involves all participants. Further "congresses" on each level would serve to interpret the separate problem areas to each other. Some of this can be done by existing governmental groups in the city or state. However, new integrative and collaborative groups must continue to evolve.

One of the continuing dilemmas will be the relationship between the various levels and forms of "authority." Clearly, those local and problem areas with the most competent participants would normally continue to get the best allocation of funds. Thus the authority on the metropolitan level would have to be responsible for seeing that funds were allocated to various problem needs and that those groups without skills or resources were given assistance to raise their levels of competence.

The function of the "metropolitan authorities" will be to catalyze and facilitate the development of programs on all levels and to connect to other "authorities"—not to control programs. Controls descending from the state or

---

[6] A discussion of a citywide forum for mental health services is being prepared by Robert Leopold, M.D., West Philadelphia Mental Health Consortium, and the University of Pennsylvania, and will be available in print in the near future.

[7] Hospital planning (especially well-documented in the Pittsburgh area, by Robert Sigmund, now at the Albert Einstein Medical Center in Philadelphia) is a good example of this evolving process.

the federal government will be unacceptable. However, when local "authorities" create standards for themselves, these standards can gradually emerge into metropolitan, and ultimately statewide or national, performance specifications. On the other hand, there must be national performance criteria, which the federal government must implement if we are to deal with inequities in our society. Too often local standards have ignored federal directions, which come only from national debate, the actions of Congress, and administrative leadership.

This outline of a pattern of governance has arisen out of the complex problems of health, but is not limited to health. It is not offered as a panacea, and, indeed, it is currently facing serious difficulties. Clearly, though, unless mechanisms are developed that bring the disparate groups and views together, we will have a continuing sense of impotence.

Any acceptable model of governance must involve the consumer and the public and private sector. It must be one that does not force compliance on local issues but demands compliance with national values and priorities, nor must it let ideological directives becloud pragmatic changes. If anything, we must leave open to professionals, with consumers, the development and choice of alternative models of governance within the broader democratic framework. This process will be turbulent. The essential theme must be decentralization, pluralization, and the involvement of consumers and all those affected by the program in the planning of their own future.

The key organizations for governance, albeit of low visability, will probably turn out to be the metropolitan ones. From the point of view of scale, that is the most relevant level of planning. But this will not arise de facto; one must accept the psychological need for subgroups to possess a feeling of territorial identify and one must work within the framework of this reality to design a low-profile, ad hoc, functionally alive network of government operations.

These concepts of government may appear to be embarking on uncharted waters, but in fact prototypes do exist in the health field, which may have extremely broad implications. One case history is worth mentioning briefly.

A California Case History

In the mental health program in the State of California,[8] funds used to be allocated by the state to a wide range of institutions on every level: state hospitals, community clinics, hospitals, county agencies, and so on. Fiscally

---

[8]Here I am indebted to Eric Plaut, M.D., Berkeley City Health Department and the University of California School of Public Health, for information about the California mental health program. In this discussion I draw on his mimeographed paper "Some Perspectives on the Development of the Lanterman, Petris-Short Mental Health Legislation of 1968."

oriented legislators, and their constituencies, were aware of the significant costs and poor results of long-term hospitalization of the mentally ill in state hospitals. Strong pressure from the political right, which felt that mental hospitals and psychiatry were essentially "bad," led to an attempt to break up the big mental hospitals and the strong mental health institutions. A sense of paranoia about psychiatry motivated an attempt to break up long-existing sets of organizational relationships. The political left wing was concerned over the deprivation of civil liberties as a result of involuntary hospitalization and long-term detention without legal review. Concurrently, there was also professional, psychiatric, and political pressure toward decentralization and the creation of local community mental health programs; and this was encouraged and supported by new federal programs. Big mental hospitals were thought to be anachronisms offering poor treatment; it was believed that most mental health care should take place in local communities close to home; and it was felt that involuntary hospitalization should be made more difficult.

Leaders in the state legislature, responding to many apparently opposing positions, were able nonetheless to create a coalition of diverse forces; as a result a new system was developed. Within the last 18 months local "mental health authorities" in the various counties throughout the state have become the focal point of all activities. The legislation provides for funding, whether from federal or state sources, to go directly to the local authorities, which contribute 10 percent of the costs but have full control over the allocation of resources. They decide whether money goes to state hospitals, private psychiatric contractors, preventive care, or their own clinics, though they must meet federal performance specifications wherever federal funds are present. New patterns of early community voluntary diagnosis were created to deal with the questions of civil liberties.

New statewide performance specifications are emerging from a "congress" of delegates from all the community mental health programs representing specific problem areas (such as juvenile delinquency, child psychiatry, alcoholism) as well as the geographically oriented "mental health authorities." This "congress" negotiates between the various locally derived performance specifications and with the State Department of Mental Hygiene. The relationship of the State Director of Mental Hygiene and the local representatives is gradually approaching the point of a "double veto." Each can veto the other's decisions. However, when mutually agreed upon, the performance specifications, which have arisen in the localities, are implemented by the state program. At the same time, the state program staff is capable of giving leadership and guidance to the local "authorities."

The local "authorities" have on the whole chosen to keep their funds within their political jurisdictions, and are increasingly spending less money on patient care in the distant state hospitals. They are choosing more and more to spend money on local programs. They are also encouraged to develop multiple funding sources and create multicounty consortia (metropolitan governances).

Here we see a picture of local control within a broader context. In the predictable future, the various local authorities will increasingly find that they need to come together in interauthority multicounty relationships on a metropolitan and regional level. These metropolitan arrangements will of necessity—and pragmatically—become the significant organizations in the recruiting of, and allocation of, resources.

Though mental health may be leading the way in the state, California is developing similar ideas for general health care. It is proposing to develop broad performance specifications for the expenditure of state and federal total health insurance; local authorities would again control the allocation of health resources, as previously outlined. The fact that the state is willing to create new mechanisms of governance and financing in health undoubtedly suggests that these solutions will serve as guideposts to more general problems in other areas.

Franchising

In the many attempts at innovation another concept is emerging which has not been seriously discussed. This is franchising, which, as it has emerged in the business community, is a way of coping with centralized performance specifications and network management while allowing for decentralized control over individual business units. Why cannot the new agents of governance serve as franchisers in other fields? Why cannot governments or public-private "authorities" offer franchises to both the public and private sectors to run a variety of services if they fit into the overall system? The beginnings of this are already evident in health, although the word franchise is not used. *Is metropolitan governance the potential major franchiser for urban programs?*

There are many arguments for the franchise concept. Moreover, it meets the psychological needs of man to have control over his immediate environment and to reject strong external controls unless these really help him. This form of organizational arrangement with metropolitan governance may well be an important part of our future.

Conclusions

We predict that vast changes are in store for us. They cover every aspect of government from the local to the federal. The specific issues of this paper focus on the implications for the metropolitan area. Clearly one of the central questions in designing governance is the question of scale, but we are faced with value questions as well. What problems can best be dealt with on a metropolitan level? What should be encouraged, supported, and catalyzed by metropolitan area agencies? Can we create this system and permit autonomy, difference, and even deviance to exist? Can we effect change that meets the imperatives of our time without total disruption and disorganization? Can human values supersede economic and technical ones? Can we extend the scope of democracy without the threat of fascism or anarchy? These are the problems of metropolitanism; indeed, these are the problems of a democratic society in the 1970s.

This essay suggests some of the solutions that will result from the interactions among human psychological behavior, societal change, and specific problem areas. Though the health field is somewhat unique, it may now be clear that the issues raised by health needs (and the value questions raised by the youth and minority groups) are more generic to the evolution of metropolitan government than was at first apparent. It is our obligation to explore thoroughly what is happening in all urban problem areas and to see whether, as is suspected, they too have similar characteristics.

4 Residuals Management and Metropolitan Governance

EDWIN T. HAEFELE and ALLEN V. KNEESE*

The purity of the air we breathe, the quality of the water we drink, and the beauty of the landscape around us are functions, to some degree, of the kind of control we exercise over the residuals generated by our production and consumption activities. That such residuals are an inevitable result of production and consumption activities has been explicated in a recent book by Kneese and others.[1] The "management" of such residuals is, therefore, a continuing process that will be with us so long as we live in towns and cities from which residuals are sufficient in quantity and quality to overwhelm the natural assimilative capacity of the ambient air, water, and land.

It must be noted that estimates, mainly from federal sources, indicate that taking environmental quality seriously is going to cost us a lot of money. More specifically for the concerns of this paper, it is going to cost cities a lot of money. A recent compilation[2] of such estimates is that municipal expenditures between 1970 and 1975 to meet water quality standards alone may run to at least $14 billion and that an additional $34 billion would be

---

*Edwin T. Haefele is a senior research associate at RFF, and Allen V. Kneese is director of RFF's quality of the environment program.

[1] Allen V. Kneese, Robert U. Ayres, and Ralph C. d'Arge, *Economics and the Environment: A Materials Balance Approach* (Washington, D.C.: Resources for the Future, Inc., 1970).

[2] Jane Brashares, "Cost Estimates for Environmental Improvement Programs," Appendix B of Allen V. Kneese, "The Economics of Environmental Pollution in the United States," in Kneese, Rolfe, and Harned (eds.), *Managing the Environment: International Economic Cooperation for Pollution Control*, Praeger Special Studies (New York: Praeger Publishers for the Atlantic Council of the United States and the Battelle Memorial Institute, December 1971).

needed if storm sewers were to be separated from sanitary sewers. Municipal expenditures for solid waste disposal, now running about $3.5 billion annually, are expected to rise by about $1 billion over the next few years. Expenditures needed in eighty-five cities to improve air quality by 5 to 15 percent with respect only to sulfur oxides and particulates are estimated to be at least $0.6 billion annually.

Estimates of the benefits of making such expenditures are much harder to come by because of estimation difficulties when dollar figures are appropriate and because of the many areas in which the political process is the direct arbiter of what the benefits are.[3] Nevertheless, some fairly hard data on the health costs of air pollution have been calculated. Lave and Seskin have estimated recently that a reduction of 50 percent in the air pollution levels in major urban areas would save over $2 billion annually in medical care and associated costs.[4]

Governments, of course, have other weapons in their arsenals than expenditures of money, and there is growing evidence that regulation of emissions, levying of user or effluent charges, and prohibitions of one kind or another will be used increasingly by municipal governments in their response to pressures for environmental improvement. A press report[5] of the proposed revision of the New York City Air Pollution Code lists the following items: phasing out the use of lead in gasoline, controlling nitrogen oxide emissions from power plants, limiting the volatility of gasoline, controlling the construction of new parking garages, limiting the sulfur contents of fuels, and outlawing the spraying of asbestos compounds. Other cities have either banned or are considering banning disposable bottles and washing compounds containing phosphates. The federal government has also passed stiff regulations on auto emissions and the use of lead in gasoline, and it is being urged to prohibit—or severely restrict—use of materials considered harmful to the environment.

All in all, there appears to be a tremendous concern about the residuals part of "residuals management" but not much concern about the "management" part. This is not said in a patronizing sense or to indicate a criticism of the administrative competence of federal, state, or local officials, but to point to two crucial facts. First, the materials-balance approach to residuals man-

---

[3] Aggregate estimates of costs and benefits, particularly those made by federal agencies that have a vested interest in one or another program, should not be taken as much more than indications of political strengths or weaknesses. As such, they indicate that federal bureaucrats and their politically sensitive masters feel that environmental improvement has some considerable potential as a source of programs, jobs, influence, and career making.

[4] Lester B. Lave and Eugene P. Seskin, "Air Pollution and Human Health," *Science* 169 (August 21, 1970): 723-33.

[5] *Environmental Science and Technology* 4 (November 1970): 882.

agement, as developed by Kneese and others, emphasizes the interdependence of gaseous, liquid, and solid residuals. Suppressing one emission in one form in one place may well create a greater problem in another form in another place. If we are to manage residuals, therefore, we must take account of the physical interdependencies. Some of the interdependencies are commonplace and well-known. For example, hauling garbage and sludge from cities to coastal waters for dumping exchanges one problem for another. It is easy to see why such practices appeal to coastal cities. The benefits of a "cheap" sink for solid waste are tangible, immediate, and desperately needed by the city, hard-pressed to meet rising welfare and education costs with a shrinking or static tax base. The costs of using the ocean as a sink are as yet unknown, the incidence of the costs are diffused, and few individuals or groups have perceived a large enough impact to resist this use of the ocean.

The ocean, the air mantle, and many of the waterways of the nation are common property resources—i.e., they are not "owned" by anybody. The land, on the other hand, is almost all in private hands (or under a specific governmental jurisdiction if in public hands). Thus it is that cities find it harder and harder to locate sites for solid waste disposal but did not until recently encounter much resistance to incineration or disposal of wastes in liquid form. Similar cases of mismanagement occur whenever one common property resource (a river, for example) comes under public management while others (the oceans, for example) are still treated as free goods.

The second fact, therefore, has to do with the inconsistencies of governmental actions about residuals. These inconsistencies are of two types: local jurisdictions are free to do things harmful to other jurisdictions and, paradoxically, they are prevented (by higher governmental restrictions, prescriptions, and proscriptions) from doing other things that would be helpful. For example, federal subsidies to municipalities for sewage-treatment plants are very likely foreclosing the possibility of scale economies in treatment-plant construction and operation. Legislation that calls for percentage cutbacks on emissions from all plants is almost certain to be a costly way to improve air quality.

The prescriptive measures taken by state and federal agencies should be recognized for what they are—compromise measures on which a minimum political consensus could be achieved. Since they deal with a large area containing many watersheds, many airsheds, and more than one population cluster, it is almost impossible for such prescriptions to reflect the interdependencies spoken of earlier, or indeed any principles of environmental management. A crude national equity is about all that can be achieved—an equity that in fact costs everybody more money than it need cost.

A metropolitan area, particularly a Standard Metropolitan Statistical Area (SMSA) that includes a fairly large hinterland, could be considered an appropriate region for residuals management if some of the restrictions from higher governments were relaxed. Trade-offs among gaseous, liquid, and solid waste alternatives could be calculated. The assimilative capacities of the air, water, and land could be assessed in relation to the residuals load. Industrial, municipal, and individual decisions could be determined and in large measure controlled.[6] The political forms by which these assessments would be made and controls decided upon are, however, crucial and there is, at present, vast confusion about political forms. The confusion manifests itself in practical arguments over city-county consolidation, metro government, special districts, local independence, and the like, but the confusion is, in reality, a confusion in theory.

Residuals Management and Social Choice

The essence of a political entity is that it possess a means of making collective (social) choices. This is true regardless of what choices are left to individual (market) choices (unless all of them are) and which are deemed collective. So long as some important decisions are not made on an individual basis, then some machinery must be set up to make them collectively. Air quality is a good example of a decision that must be made collectively if it is made at all. (We explore why in a later section.) A number of problems are faced as air quality becomes a social-choice issue, and they illustrate the theoretical issues involved. They may be posed as follows:

(1) How large an area is to be affected by the decision on air quality?
(2) What air quality level(s) should be established?
(3) How should the expense of attaining these levels be borne?
(4) By whom should the decision be made?
(5) By what means should the air quality levels be achieved?

The answers to these questions may be constrained by technological limits (production possibilities), by economic limits, and, in practice, by political and financial limitations.[7] Although these limitations will dominate the problem in most specific situations, underlying them is a more fundamental issue. What theory or normative model could be used to guide our answers?

The last question addresses the issues of "participatory" democracy, economic efficiency, the responsiveness of the system, minority representation and control, income redistribution, community values in planning, and the

---

[6] Analytical models to serve these purposes are under rapid development. See the paper by Clifford S. Russell and Walter O. Spofford, Jr., in Allen V. Kneese and Blair T. Bower (eds.), *Environmental Quality Analysis: Research Studies in the Social Sciences* (Baltimore: Johns Hopkins Press, 1972).

[7] They are, most certainly, influenced by the trade-offs among residuals—i.e., higher air quality levels may be achieved at the expense of greater problems (lower quality) in water.

other value conflicts about goals and methods. The questions are especially hard to discuss because of the wide disparity of views, most held subconsciously, about them. Nevertheless it is essential to make the attempt.

In an attempt to constrict the area of disagreement, let us assume that everyone would agree that decisions cannot be made in the absence of information about the following:

(1) Present air quality at various locations in terms of sulfur oxides, particulates, and so forth.

(2) What residuals are now being discharged into the atmosphere at what points.

(3) Alternative methods of achieving reductions of ambient concentrations and the costs (including employment effects) of such reductions.

There will be some anti-system people unwilling to allow even that degree of information collecting, but ignoring them, we have already established a need both for a monitoring function and for some technical analysis of alternative actions for improving air quality.

If we now admit the interdependence of physical forms of residuals, we cannot resist monitoring and analyzing alternative costs and impacts on waterborne wastes and solid wastes. In other words, regardless of how we decide to handle the questions of quality goals, extent of coverage, financing, and the machinery for choosing; *for air*, we have committed ourselves to fairly elaborate residuals monitoring and analysis functions for the air, water, and solid waste of some region. If we are to be rational, there is no escaping the obligation to perform these functions.

Keep in mind, however, that we have not committed ourselves to any particular way of performing them. We may have a civil service perform them; we may contract with a private company; we may create a public corporation. All three are in fact done.

We now are contemplating a metropolitanwide (at least) monitoring and analysis system for gaseous, liquid, and solid wastes. We have arrived here by taking the obvious and least controversial options. We are not far from the view of most experts in the field. Still, in our attitude of suspended disbelief, let us ask this monitoring and analysis system to perform on a metropolitan (at least) basis and look at the results. The results could, of course, be transmitted to the state and/or to each local jurisdiction in the area. No specific governmental form is necessary for this much of our system to work, just a little information cost-sharing. It is worth calling attention to this fact, because many assume that a metropolitan government, or metropolitan special authority or district, "must" be set up in order to get this information. Moreover, it is even possible that voluntary, unanimous agreement by all municipalities in the area could result in the establishment of a program based on information gathered through such a monitoring and analysis system and that enforcement could similarly be agreed upon and undertaken, even on a contract basis. Councils of government can exist and function by this Wicksellian machinery.

If we wish to move beyond the strictures of the unanimity rule and councils of government, we face some serious theoretical and practical questions, the answers to which are far from clear. Our monitoring and analysis service may tell us, in some detail, how we may efficiently raise air quality (in a total residuals management context) by various increments. The impacts and benefits of such moves will probably vary over the territory involved. How we move on the decision now makes an enormous difference. If we ask municipalities, the response will probably be different from the response of county governments. If we ask interest groups, the response might be still different. If we imagine equal population wards, the answer is again different. Moreover, a referendum vote will produce a different answer from that worked out in a legislative process of representatives in the same jurisdiction. A survey, conducted for a planning commission, might produce another answer. A hearing would generate its own unique confusion.

It is important to stress that the answers would not be different simply because of inefficient information-gathering processes. The answers are different because the way the questions would be asked, the way the issues would be formed, and the process of resolution or decision by the different units would all be different.

Whom do you ask? What proposal(s) do you ask about?

Framing the issues for decision and deciding upon the appropriate collectivity to ask is the essence of the government problem. It is at the heart of the value conflict problem. It is at the heart of the welfare economics problem with a social welfare function. It is the nexus of confusion in much social science theory.

Haefele has attempted elsewhere[8] to provide a way out of the thicket through a utility analysis. He showed that individual values could be "appropriately" related to social choices through a two-party system of representative government. The criterion for "appropriate" was that the government choose the same policy as would be chosen if everyone were in an assembly and vote-trading on the issues were allowed. Representative government, under a two-party system competing for votes in single-member districts, proved to be a mechanism capable of producing solutions identical to those chosen by direct voter trades.[9]

---

[8] Edwin T. Haefele, "A Utility Theory of Representative Government," *American Economic Review* 61 (June 1971): 350-67.

[9] This statement should not be taken to be an assertion that present, existing governmental forms or parties are capable of producing this "ideal" outcome any more than present businesses sell at competitive equilibrium prices.

The issue of how to choose the collectivity, that is to say, what boundaries to use, is yet to be faced. Madison suggested that the rule for boundary setting for governments of general jurisdiction was to encompass a heterogeneous population with common problems. Applying that rule still leaves us with a large number of possible choices. The physical interdependencies in residuals management, referred to previously, help to narrow the choices. For example, the economic reach of a metropolitan area, its economic base, is reasonably consonant with residuals management problems associated with air quality and solid waste. Moreover, the economic reach is a useful one for considering a tax base, a transportation system, and land-use controls.

An entire watershed is, however, rarely encompassed by a metropolitan region, and water quality management is most efficiently done on a watershed basis. Thus, problems of coordinating executive government and of setting legislative policy exist because of this noncongruence of problem-shed boundaries. The usual approach to watershed management, the interstate compact, has its own set of problems[10] but may be a useful solution if state lines are reasonably consonant with the watershed. When they are not, a separate executive agency for watershed management, controlled by a board composed of the concerned metropolitan regions, may be the most easily arranged way to effect policy coordination while realizing the efficiency gains possible from basinwide management of water resources.

In summary, although we cannot completely resolve the boundary question, we suggest that a new government is most apt to be successful at the metropolitan-region level (where a number of common problems appear to coexist and the population is heterogeneous) and that several options exist to make the necessary ties between that general government and the watershed. We are led to this conclusion in part because the legislative mechanism for making social choices—vote-trading—can function more efficiently when many independent issues (transportation, land use, tax rates, air quality) arise in the same territory,[11] and in part because the residuals management (executive mechanism) function can be carried on efficiently at this level.

It may be worth remarking that existing states might qualify on most of the points made. Are we just re-inventing state government? That we are not is emphasized by a recent action of the Maryland legislature regarding a Washington Metro (transit) bill.[12] Regarded as a "local" issue (not a common problem), the vote-trading necessary on the part of the Washington area

---

[10] See E. T. Haefele, "Environmental Quality as a Problem of Social Choice," in Kneese and Bower (eds.), *Environmental Quality Analysis*.

[11] See "Coalitions, Minority Representation and Vote-Trading Probabilities," *Public Choice*, Spring 1970, pp. 75-90.

[12] Reported by the *Washington Post*, April 10, 1971, under the heading, "The Metro Bill and the Sea Full of Sharks."

representatives was far beyond that which would have been required in a metropolitan legislature. The reason: since most state legislators had *no* interest one way or the other, intensities of preferences, which are the essence of vote-trading, played no part. In effect, free money was introduced into the system, driving up prices to all who *were* interested in the outcome. On a metropolitan level (where Metro is a common problem), all representatives would have *some* interest in the transit bill and the intensity of those interests could be appropriately expressed through vote-trading.

It is also worth noting that whereas a legislature composed of, say, equal population districts within a metropolitan region provides a means of aggregating individual preferences, *given the issues*, it is quite likely to be parochial as a generator of issues. Thus, some provision should be made either for some representation at large in the proposed legislature or for an elected executive with veto power. The former ensures that metropolitanwide interests get introduced into the deliberations; the latter goes further and ensures that these interests must concur on any action undertaken.

Collective Choices and Common Property Resources

The rudimentary and ghostly forms of executive and legislative bodies we have hypothesized are sufficient to confront a major practical problem. Again, although it is practical, the reason it is a problem is that there is some confusion in the theory.

The quickest way to get into the problem is to say that the analysis section (executive) of our hypothesized government will be perfectly willing to make policy de facto, using their own estimate of political reality, and the legislators will be tickled to death to let them do so, under the pretense that they pursued an objective, scientific method of decision making by which costs and benefits were appropriately measured, citizen and special group interests "taken into account," and the final choice (albeit clearly signposted) actually "made" by the legislature. Since most of the readers of this paper will recognize this situation and have their own reactions to it (ranging from dismay to joyful acceptance), we shall not dwell on a description of a situation that has put most of the real choices in executive hands, relegated the legislature to a legitimating role, and allowed many legislators to be all things to all people.

The purpose of this section is to argue that, failing a return of royal prerogative as the basis, there is no way the executive branch can make the collective choices consonant with the theory of representative government and no way that collective choices can be avoided in decisions about resources that are common property.

The first point can be disposed of quickly. Even though a mayor, governor, or other executive official is elected, he is elected by *one* majority. It is almost a truism of representative government in America that the only defense against "the tyranny of the majority" is to lodge policy determination in the legislature, where different issues bring forth different majorities and intensities of preferences can be used (vote-trading) to arrive at final decisions. The only analogue to this process in the executive is the election strategy by which a candidate constructs a bundle of issues and positions to run on. One could imagine the two processes coming to the same conclusion, but it requires the imagination of a socialist planner who says he can come to the same equilibrium solution as a free market system does. It is an interesting, though pathetic, commentary on our present predicament that cyberneticists[13] now try to replace representative government by a system of feedback loops from executive policies to citizen reactions to new executive policies. This approach, which rejects the distinction between legislative and executive functions, puts us back to the organic state of Tudor times and ignores the constitutional battles of the seventeenth century. But, since American constitutions, federal and state, are the direct results of those seventeenth century battles, such approaches create a paralysis of government. The paralysis occurs because of attempts to force executive governments to do what constitutional machinery prevents them doing—to make collective choices. The constitutional bar, a result of political history, is valid on technical grounds —executive government choices fail to converge (the historical demonstration occurs convincingly in the Protectorate), and they are not stable (since trade-offs can neither be discovered nor efficiently carried out except in the most coarse-grained sense. The reign of James II illustrates this failure.)

The second point, the necessity for collective choices about common-property resources, must be raised if only because of attempts by some to force all such questions back to marketplace or pseudo-marketplace solutions. The incentive to do so is, of course, great. Economic theory suggests that individual decision making about what kinds and amounts of goods and services to buy will lead, through a market exchange system, to the optimal level of production of these goods and services, given certain assumptions about ideal markets, and lack of externalities (third-party effects). Such theory proceeds from a given distribution of income and assumes that all goods and services can be divided at the individual level—i.e., are not consumed in common.

For goods and services not capable of being so treated, a more complicated theory is needed. Specifically, we must consider distributional criteria as well as efficiency criteria. The income-distribution problem associated with these "public goods" is particularly difficult because the consumption of the good cannot be differentiated among consumers on the basis of their voluntary

---

[13] See E. S. Savas, "Cybernetics in City Hall," *Science* 168: 1066-71.

choice in markets. When the supply of a public good changes, both efficiency and distribution are affected, and there is, in general, no way to be sure that equating marginal cost with the sum of marginal willingness to pay will be a welfare maximum. This has been a hard problem for applied public economics, and several devices have been used to get around it. Otto Eckstein, in his work on water development,[14] explicitly assumed that the marginal utility of income (a cardinal measure not employed in contemporary analysis) is the same for all individuals. This effectively wipes out distributional considerations, but most economists regard it as grossly unrealistic.[15]

The assumption most often made implicitly by practicing economists is that it is a mistake to consider individual public goods as situations in isolation. Rather the whole complex of public goods should be considered. Some will affect one group adversely and another favorably and others vice versa. Thus there will be a lot of cancellation of distributional effects. The society that makes its decisions based on efficiency criteria will be one in which most people will finally be better off than one in which criteria are used that foreclose efficient solutions. It is further assumed that public goods are a rather small part of the economy and that private goods are allocated (through tax and subsidy policy) in an ethically sanctioned way.

A second, but separate, problem has to do with the measurement of willingness-to-pay. Samuelson[16] has argued that the willingness to pay for public goods cannot be measured because exclusion is not possible. This is a frequently repeated fallacy in the literature. It results from the fact that most theoretical discussions start from the idea that one would ask people what they would be willing to pay. In practice this technique is hardly ever used. It is also not correct to say that persons cannot be excluded from the provision of public goods. Frequently exclusion is *possible*; but if we are talking about a pure public good, it is *undesirable* on efficiency grounds. Exclusion arguments have done much to muddy the waters. There is a continuum of exclusion problems with most public goods near one end and most private goods near the other. It is nevertheless true that markets do not provide *direct* information on the value of public goods and, therefore, estimation problems

[14] Otto Eckstein, *Water Resource Development: The Economics of Project Evaluation* (Cambridge, Mass.: Harvard University Press, 1961).

[15] Another device is to assume $\delta s^i/\delta y^i$ (where δs^i is the marginal rate of substitution of private goods for the public good of individual i, and δy^i the marginal change in income of individual i) is the same for all individuals i, then one can separate the distribution problem from the allocation problem. That is, different distributions will not affect the optimal allocation. In applied ordinal models this assumption is necessary for consistent aggregations of consumer surpluses. Karl Göran Mäler has proved this point in an unpublished RFF manuscript.

[16] "The Pure Theory of Public Expenditures," *The Collected Scientific Papers of Paul A. Samuelson*, ed. Joseph E. Stiglitz (Cambridge, Mass.: M.I.T. Press, 1966), p. 1223.

are always difficult and sometimes close to impossible. Accordingly, even when strong efforts are made to obtain willingness-to-pay information on the range of public good candidates, it is likely to be partial and of widely differing dependability from case to case.

There has been much work on willingness-to-pay in the context of common property resources, primarily because the advantage of using the price route is that the allocation is then automatic and need not take anyone's time (beyond a policing or enforcement capability). The advantage is significant. A further advantage is that revenues are generated, and no better source of revenues could be imagined than charges on effluents. A third advantage of prices is that they can be efficient—encouraging firms, municipalities, and individuals to seek lower-cost methods of dealing with their residuals through changes in their amount, timing, and physical forms.

The mechanism of price, as such, is unexceptionable. The level of prices is another matter. We can easily agree that relative prices should depend on relative damage caused to the resources or to the relative costs to process the effluent. The *level* of prices, on the other hand, relates to a quality specification. What quality air shall we have? Can we find *that* out by estimating people's willingness to pay? Let us explore this idea under the best possible assumptions, i.e., by ignoring the distribution-of-income problem, by ignoring the estimation problem, and by assuming that the externalities issue is incorporated in the pricing schedule and thus can also be ignored. In other words, if income redistribution is set aside, if we possessed schedules of people's willingness to pay (setting aside the estimation problem), and if our pricing scheme accounted for all external effects of each person's decisions, could we determine the appropriate level of air quality directly by measuring willingness-to-pay without asking for a collective choice on the matter?

Clearly, under this highly artificial set of assumptions, we could equate marginal cost to the sum of marginal willingness-to-pay. We could, therefore, provide the "optimal" level of air quality, i.e., that level beyond which no one is willing to bear the next increment of cost. But, since there are physical links between air, water, and solid residuals, we should expand our willingness-to-pay schedule to encompass liquid and solids as well as gaseous residuals in order to arrive at environmental willingness-to-pay schedules, on the basis of which the equating or marginal costs could be accomplished. Moreover, since environmental quality is only one of many publicly provided goods and services, we cannot ignore the interrelationships with these other goods and services.

In short, a partial solution based on willingness-to-pay will not suffice in an equilibrium setting. We should have to have willingness-to-pay schedules on all publicly provided goods. It may be wholly appropriate to use willingness-to-pay data in a recreation project analysis as a guide to investment in *one* park, lake, or whatever. This is a legitimate partial analysis in which executive government, under some policy authorization and budget constraint, is attempting to discover (much as a private firm would) where the payoff is highest. It is a totally different thing to use a willingness-to-pay measure to ascertain how many parks or lakes an area should have (or what level of air quality). The latter questions are better asked by political entrepreneurs who are responsible for general taxation if intensities of preferences (consumer surpluses) about air quality are to be used to accommodate policies or investments in other areas where there are intense preferences also. Assessing willingness-to-pay may, however, play a very useful role in helping to focus the political process, particularly when it can be shown that large welfare gains are possible. Such evidence gives incentive to political entrepreneurs and ammunition to the public. The resulting collective choice, which provides the simultaneous solution to the efficiency and income questions, benefits from both. Indeed, it may be that estimating willingness-to-pay and providing other measures of benefit and cost (including incidence of such benefits and costs) are especially useful as issue formation devices for the process of collective choice.

Conclusion

Starting with the problem of choosing a level of air quality in a metropolitan area, we have inductively concluded that an answer depends on a collective choice by a "regional" general-purpose government. We arrive there not only because of scale economies in the production possibilities for air quality but also because of some hard problems concerning common property resources, public goods, and the theory of collective choice. To come to such a commonplace destination by such a tortuous route has two possible implications—either the answer has a fair chance of being right, or the authors started out with the conclusion already in mind.

Without pursuing either implication, there remain some additional points to be made about this conclusion. First, the fact that individuals can move around in a metropolitan area choosing that mix of public services (including air quality) that comes closest to their own preferences does not obviate the need for the collective choice at the "regional" level. Inasmuch as no single municipality can control its air quality by its own ordinances, some overall agreement is essential. That agreement simply defines the range (not merely the lower bound) over which air quality will vary from place to place in the metropolitan area. It neither replaces individual choice (exercised by moving) nor hinders it.

Second, we have not specified the executive arm of our "regional" government. It is entirely possible that a public corporation could function as such.

The new Maryland Environmental Service Agency, although not an exact parallel, is a useful analogue. Operating at the state level under the direct policy control of the state legislature, the MES can contract with any municipality or industry to handle residuals and may construct and operate treatment plants. Its charges reflect its costs, as it must be self-sufficient. Since no industry or municipality is obligated to use ESA services, it must be efficient to survive.

Third, we have not specified the legislative arm of our "regional" government, beyond the prescription that it be an elected body from single-member equal population districts. The closest analogue here is the Metropolitan Council of Minneapolis-St. Paul, a newly created government of broad jurisdiction that exercises policy control over park boards, aviation commissions, sewage districts, and the like. Proposals are in hand for the Council to become an elected body (members are now appointed by the Governor from equal population districts) and for a metropolitanwide tax-sharing scheme.

Fourth, we have not defined the limits of our region with much precision. We have shown that the technical interdependencies of residuals management imply an inner boundary—the metropolitan area or economic region. We have indicated that the outer bound may well be less than a state (although obviously when the economic region crosses state lines the region will comprise parts of two or more states). These limits are unsatisfactory, however, because some economic areas (the New York metropolitan area, for example) are larger than some states and all the problems of the area are not common (though the residents are heterogeneous enough). The boundary problem awaits further work after all pieces of the governance problem have been investigated. It cannot be solved by the perspective of one problem.

Finally, we have suggested a legislative body as a goal-setting mechanism and a pricing scheme (effluent charges) as the means of implementing the goals, conscious that neither will function perfectly when in place. Approximations of marginal damage, estimations of effluent effects, and so forth will abound; yet the pricing mechanism remains superior to any other feasible means because it allows each firm, person, and municipality to make individual decisions that reflect social costs. Similarly, a look at existing legislatures is enough to show that their defects are pervasive. Yet, again, the legislature remains the only governmental institution capable of making social choices if we value individual preferences as creators of social choice. In short, because both instruments have the capacity for theoretical perfection, we are able to evaluate their shortcomings in practice and, over time, to correct them.

For Product Safety Concerns and Information please contact our EU representative GPSR@taylorandfrancis.com
Taylor & Francis Verlag GmbH, Kaufingerstraße 24, 80331 München, Germany

www.ingramcontent.com/pod-product-compliance
Lightning Source LLC
Chambersburg PA
CBHW050541270326
41926CB00015B/3326